Instructor's Manual for

CHEMISTRY

Principles & Reactions

SECOND EDITION

William L. Masterton

University of Connecticut

SAUNDERS GOLDEN SUNBURST SERIES

SAUNDERS COLLEGE PUBLISHING

Fort Worth • Philadelphia • San Diego • New York • Orlando • Austin • San Antonio

Toronto • Montreal • London • Sydney • Tokyo

Printed in the United States of America.

Masterton: Instructor's Manual to accompany
CHEMISTRY: PRINCIPLES & REACTIONS, 2/E

ISBN 0-03-028997-1

345 021 98765432

PREFACE

This manual starts off with a section entitled "Lecture Outlines", which you may find helpful in adapting the text to your class schedule. Beyond that, the manual is organized by text chapters. For each chapter, we include four different features:

1. "Lecture Notes", which indicate the amount of time that we devote to each chapter and the topics that we emphasize. Included are detailed lecture outlines, which may serve as a guide for your own lectures. At a minimum, they indicate how we cover topics and how successive topics can be integrated.

2. A list of demonstrations illustrating topics covered in the chapter. These are taken from four sources:

- the manual "Tested Demonstrations in Chemistry" (6th edition) compiled by Alyea and Dutton and published by the Journal of Chemical Education in 1965. These are coded as "Test. Dem." with the page numbers.

- demonstations described in the Journal of Chemical Education between 1965 and 1992. Here, the journal reference is given.

- "Chemical Demonstrations", Vol. 1-4, published by Bassam Shakhashiri with many collaborators and contributors. These are listed as "Shak." followed by the appropriate volume and page reference.

- Shakhashiri Videotapes, many of which illustrate demonstrations described in "Chemical Demonstrations" by the same author. Here the demonstration number is cited.

3. Quizzes suitable for use in discussion sections. Each quiz should take 10-15 minutes. There are five quizzes for each chapter. The questions included could be used for examinations if you prefer.

4. Answers and detailed solutions to

- the text problems which do not have answers in Appendix 4.

- the challenge problems at the end of each chapter.

Note that the problems in the text that are numbered in color are answered in Appendix 4. Detailed solutions to these problems are given in the "Solutions Manual", written by Professor John Bauman and available from the publisher.

TABLE OF CONTENTS

LECTURE OUTLINES

This text, unlike other general chemistry texts now available, can be covered in its entirety in a one-year course. A reasonable schedule is shown below; further comments on the amount of time you should devote to each chapter are given in the body of this manual. It is assumed for convenience that you are teaching 14-week semesters with two 50-minute lectures per week. If three class periods are devoted to examinations each semester, that leaves 25 for covering material. On that basis, you should be able to complete Chapter 10 (Solutions) in the first semester. The second semester will then start with Reaction Rates (Chapter 11).

FIRST SEMESTER SCHEDULE

Week	Lecture	Topic
1	1	Chapter 1 (Matter and Measurements)
	2	Chapter 1
2	3	Chapter 2 (Atoms, Molecules, and Ions)
	4	Chapter 2
3	5	Chapter 3 (Mass Relations in Chemistry)
	6	Chapter 3
4	7	Chapter 3
	8	EXAM I
5	9	Chapter 4 (Reactions in Aqueous Solution)
	10	Chapter 4
6	11	Chapter 4
	12	Chapter 5 (Gases)
7	13	Chapter 5
	14	Chapter 6 (Electronic Structure)
8	15	Chapter 6
	16	Chapter 6
9	17	EXAM II
	18	Chapter 7 (Covalent Bonding)
10	19	Chapter 7
	20	Chapter 7
11	21	Chapter 8 (Thermochemistry)
	22	Chapter 8
12	23	Chapter 9 (Liquids and Solids)
	24	Chapter 9
13	25	Chapter 9
	26	EXAM III
14	27	Chapter 10 (Solutions)
	28	Chapter 10

SECOND SEMESTER SCHEDULE

Week	Lecture	Topic
1	1	Chapter 11 (Rate of Reaction)
	2	Chapter 11
2	3	Chapter 11
	4	Chapter 12 (Gaseous Chemical Equilibrium)
3	5	Chapter 12
	6	Chapter 13 (Acids and Bases)
4	7	Chapter 13
	8	EXAM I
5	9	Chapter 13
	10	Chapter 14 (Acid-Base and Precipitation Equil)
6	11	Chapter 14
	12	Chapter 15 (Complex Ions)
7	13	Chapter 15
	14	Chapter 16 (Spontaneity of Reaction)
8	15	Chapter 16
	16	EXAM II
9	17	Chapter 17 (Electrochemistry)
	18	Chapter 17
10	19	Chapter 17
	20	Chapter 18 (Nuclear Reactions)
11	21	Chapter 18
	22	Chapter 19 (Chemistry of the Metals)
12	23	Chapter 19
	24	EXAM III
13	25	Chapter 20 (Chemistry of the Nonmetals)
	26	Chapter 20
14	27	Chapter 21 (Organic Chemistry)
	28	Chapter 21

If you want to use lecture time for review, for going over assigned problems, or for doing a large number of demonstrations, you will have trouble keeping up with this schedule. As you've almost certainly learned by now, the solution to this problem is not to talk faster; judicious deletions work better. It's been said, and wisely, that the secret of giving a good lecture is knowing what to leave out. Possible candidates include:

- introductory material on matter in Chapter 1 and atomic theory in Chapter 2. The chances are your students have been exposed to this material more than once in high school and understood it reasonably well the first time.
- Boyle's and Charles' laws in Chapter 5. We start the chapter by writing the ideal gas law and go on from there
- the First Law discussion in Chapter 8. Quite frankly, this has very little to do with chemistry; students will not be irreparably damaged if they are unaware of the distinction between H and E.
- the crystal structure, unit cell discussion in Chapter 9. We cover this in class but can easily understand that other instructors might not care to do so. Along the same line, the

discussion of colligative properties in Chapter 10 could probably be drastically shortened. In particular, Raoult's law could easily be omitted.

- reaction mechanisms in Chapter 11. Students have a lot of trouble with this; we're not sure it's worth the effort.
- nomenclature of complex ions in Chapter 15, which is not one of the more exciting topics in general chemistry
- nuclear structure in Chapter 18
- qualitative analysis and/or alloys in Chapter 19

Finally, we should point out that many instructors will want to spend more time on descriptive inorganic and/or organic chemistry (Chapters 19-21). Others may want to cover in class the material in Appendix 5 on net ionic equations; it's a good way to tie together the year course in general chemistry. If you decide to move in either of these directions, you'll have to make some deletions from the schedules discussed above. Please don't omit Chapter 15 (Complex Ions) or Chapter 18 (Nuclear Chemistry); the students enjoy them and so do we.

CHAPTER 1
Matter and Measurement

LECTURE NOTES

This material ordinarily requires 2 lectures (100 min), allowing for a 10-15 minute introduction to the course in the first lecture. If you're in a hurry, this can be cut to 1½ lectures by discussing only quantitative material (significant figures, unit conversions, density, solubility).

A few points to keep in mind:

- virtually all of your students will be familiar with the metric system and prefixes. It may be worth discussing the basis of SI, but you don't have to dwell on it.

- students readily learn the rules of significant figures but typically ignore them after Chapter 1. It may help to emphasize that these are common-sense (albeit approximate) rules for estimating experimental error.

- many students (typically the weaker ones) stubbornly resist using conversion factors, preferring instead a rote method with which they became infected in high school. It may be useful to point out that conversion factors will be used throughout the course and hence should be learned at this point.

- students often have trouble with solubility calculations. The approach used in the text involves conversions (Example 1.8). The solubility is considered to be a conversion factor relating grams of solute to grams of solvent.

LECTURE 1

I Types of Substances

A. Elements Cannot be broken down into simpler substances.
Examples: nitrogen, lead, sodium, arsenic. Symbols: N, Pb, Na, As

B. Compounds Contain two or more elements with fixed mass percents.
Sodium chloride: 39.34% Na, 60.66% Cl
Glucose: 40.00% C, 6.71% H, 53.29% O

C. Mixtures Homogeneous (solutions) vs heterogeneous. Separation by filtration, distillation.

II Measured Quantities

A. Length Base unit is the meter. 1 km = 10^3 m; 1 cm = 10^{-2} m; 1 mm = 10^{-3} m; 1 nm = 10^{-9} m. Dimensions of very tiny particles will be expressed in nanometers.

B. Mass 1 kg = 10^3 g; 1 mg = 10^{-3} g. Point out that two different kinds of balances will be used in lab. Analytical balance (±0.001 g) should be used only for accurate, quantitative work.

C. Temperature $t_{°F} = 1.8t_{°C} + 32°$; $T_K = t_{°C} + 273.15$
Convert 68°F to °C and K:
$t_{°C} = (68° - 32°)/1.8 = 20°C$; $T_K = 293$

D. Derived Units
1. Volume 1 L = 10^3 mL = 10^3 cm^3 = 10^{-3} m^3
2. Energy Joule = energy consumed when 10-watt bulb burns for 0.1 s. 4.184 J = 1 cal. Burning match evolves about 0.5 kcal = 2 kJ

LECTURE 2

I Experimental Error; Significant Figures

Suppose object is weighed on crude balance to ±0.1 g and mass is found to be 23.6 g. This quantity contains 3 "significant figures", i.e., 3 experimentally meaningful digits. With an analytical balance, mass might be 23.582 g (5 sig. fig.)

A. Counting significant figures
1. Volume of liquid = 24.0 mL; three significant figures. Zeros at end of measured quantity, following nonzero digits, are significant.
2. Volume = 0.0240 L; three significant figures (note that 0.0240 L = 24.0 mL). Zeros at beginning of a measured quantity, preceding nonzero digits, are not significant.

B. Multiplication and Division Keep only as many significant figures as there are in the least precise quantity. Density of piece of metal weighing 36.123 g with volume of 13.4 mL?

$$\text{density} = \frac{36.123\ g}{13.4\ mL} = 2.70\ g/mL$$

C. Addition and Subtraction Keep only as many digits after the decimal point as there are in the least precise quantity. Add 1.223 g of sugar to 154.5 g of coffee:

$$\text{Total mass} = 1.2\ g + 154.5\ g = 155.7\ g$$

Note that rule for addition and subtraction does not relate to significant figures. Number of significant figures often decreases upon subtraction:

Mass beaker + sample =	52.169 g	5 sig. fig.
Mass empty beaker =	52.120 g	5 sig. fig.
Mass sample =	0.049 g	2 sig. fig.

D. Exact Numbers "one liter" means 1.00000 . . . L

II Conversion Factors

A. Simple, one-step conversions

1. A rainbow trout is measured to be 16.2 in long. Length in centimeters?

$$\text{length in cm} = 16.2 \text{ in} \times \frac{2.54 \text{ cm}}{1 \text{ in}} = 41.1 \text{ cm}$$

Note cancellation of units. To convert from centimeters to inches, would use the conversion factor 1 in/2.54 cm

2. Barometric pressure reported on Canadian radio to be 99.6 kPa. Express in mm Hg (101.3 kPa = 760 mm Hg).

$$\text{pressure (mm Hg)} = 99.6 \text{ kPa} \times \frac{760 \text{ mm Hg}}{101.3 \text{ kPa}} = 747 \text{ mm Hg}$$

Method is particularly useful with unfamiliar units.

B. Multiple Conversion Factors Baseball thrown at rate of 89.6 miles per hour. Speed in meters per second?

$1 \text{ mile} = 1.609 \text{ km} = 1.609 \times 10^3 \text{ m}$; $1 \text{ h} = 3600 \text{ s}$

$$\text{speed} = 89.6 \frac{\text{mile}}{\text{hr}} \times \frac{1.609 \times 10^3 \text{ m}}{1 \text{ mile}} \times \frac{1 \text{ h}}{3600 \text{ s}} = 40.0 \text{ m/s}$$

III Properties of Substances

Distinguish between intensive vs extensive, chemical vs physical.

A. Density Empty flask weighs 22.138 g. Pipet 5.00 mL of octane into flask: total mass = 25.598 g

$d = 3.460 \text{ g}/5.00 \text{ mL} = 0.692 \text{ g/mL}$

Volume occupied by ten grams of octane?

$$V = 10.00 \text{ g} \times \frac{1 \text{ mL}}{0.692 \text{ g}} = 14.5 \text{ g}$$

B. Solubility Often expressed as grams of solute per 100 g solvent. Example:

temperature	10°C	100°C
soly. lead nitrate (g/100 g water)	50	140

1. How much water is required to dissolve 80 g of lead nitrate at 100°C?

   $$\text{Mass water} = 80 \text{ g lead nitrate} \times \frac{100 \text{ g water}}{140 \text{ g lead nitrate}}$$

   $$= 57 \text{ g water}$$

2. Cool to 10°C. How much lead nitrate remains in solution?

   $$\text{Mass lead nitrate} = 57 \text{ g water} \times \frac{50 \text{ g lead nitrate}}{100 \text{ g water}}$$

   $$= 28 \text{ g lead nitrate}$$

   80 g - 28 g = 52 g lead nitrate crystallizes

DEMONSTRATIONS

1. The scientific method: J. Chem. Educ. 66 597 (1989)
2. Reaction of iron with sulfur: Shak. 1 55
3. Chromatography: Test. Dem. 119, 137; J. Chem. Educ. 59 1042 (1982), 62 530 (1985)
4. Density: Test. Dem. 155

*5. Miscibility and density of liquids: Shak. 3 229

* See also Shakhashiri Videotapes, Demonstation 36

QUIZZES

Quiz 1

1. The density of a certain liquid is 69.4 lb/ft^3. Express this in grams per cubic centimeter, given:

   1 lb = 453.6 g   1 ft = 30.48 cm

2. The solubility of potassium nitrate at 100°C is 240 g/100 g water.
   a. How much water is required to dissolve 112 g of potassium nitrate at 100°C?
   b. How much potassium nitrate can be dissolved in 56.2 g of water at 100°C?

Quiz 2

1. Assume your car has a fuel economy of 28.2 miles per gallon. Express this in kilometers per liter, given

   1 mile = 1.609 km   1 gal = 4 qt   1 L = 1.057 qt

2. The density of methanol is 0.787 g/mL
   a. What is the mass of 25.0 mL of methanol?
   b. What volume of benzene (d = 0.879 g/mL) has the same mass as the methanol in (a)?

Quiz 3

1. An atom of silicon has a radius of 0.117 nm. Calculate its volume ($V = 4\pi r^3/3$) in:
   a. cubic nanometers b. cubic inches (1 in = 2.54 cm; 10^9 nm = 1 m)
2. Give the number of significant figures in
   a. 0.0242 mL b. 3.00 g c. 320 L

Quiz 4

1. A helium atom at room temperature is moving at an average speed of about 1.36×10^5 cm/s. Express this in miles per hour.

   1 mile = 1.609 km   1 h = 3600 s

2. A helium atom weighs 6.65×10^{-24} g and has a radius of 0.050 nm. Calculate the density of the atom in g/cm^3. ($V = 4\pi r^3/3$)

Quiz 5

1. The solubility of sodium chloride at room temperature is about 35 g/100 g water. Express the solubility in pounds of sodium chloride per liter of water.

   1 lb = 453.6 g   density water = 1.00 g/mL

2. Using the solubility quoted in (a), calculate
   a. the mass of sodium chloride that will dissolve in one kilogram of water.
   b. the mass of water required to dissolve 1.00 g of sodium chloride.

Answers

Quiz 1: 1. 1.11 g/mL 2. a. 46.7 g b. 135 g

Quiz 2: 1. 12.0 km/L 2. a. 19.7 g b. 22.4 mL

Quiz 3: 1. a. 6.71×10^{-3} nm^3 b. 4.10×10^{-25} in^3
2. a. 3 b. 3 c. 2 or 3

Quiz 4: 1. 3.04×10^3 mph 2. 13 g/cm^3

Quiz 5: 1. 0.77 lb/L 2. a. 350 g b. 2.9 g

PROBLEMS

1. a. mass b. volume c. volume d. length e. density
   f. pressure g. energy

3. a. $0.300 \text{ km} \times \frac{10^3 \text{ m}}{1 \text{ km}} = 300 \text{ m} < 303 \text{ m}$

   b. $0.0500 \text{ g} \times \frac{10^{-3} \text{ kg}}{1 \text{ g}} = 5.00 \times 10^{-5} \text{ kg} < 500 \text{ kg}$

   c. $1.50 \times 10^3 \text{ nm}^3 \times \frac{(10^{-7} \text{ cm})^3}{1 \text{ nm}^3} = 1.50 \times 10^{-18} \text{ cm}^3 < 1.50 \text{ cm}^3$

   d. $2.50 \times 10^{-3} \frac{\text{kg}}{\text{m}^3} \times \frac{10^3 \text{ g}}{1 \text{ kg}} \times \frac{1 \text{ m}^3}{10^6 \text{ cm}^3} = 2.50 \times 10^{-6} \frac{\text{g}}{\text{cm}^3} < 25.0 \frac{\text{g}}{\text{cm}^3}$

5. a. $t_{°F} = 1.8(25°) + 32° = 77°F$

   b. $T_K = 25 + 273.15 = 298 \text{ K}$

7. $-69.7° = 1.8t_{°C} + 32°; \quad t_{°C} = -56.5°C$

9. a. $1 \text{ mL} = 10^{-6} \text{ m}^3$ b. $1 \text{ J} = 1 \text{ kg}\cdot\text{m}^2/\text{s}^2$ c. $1 \text{ Pa} = 1 \text{ kg/m}\cdot\text{s}^2$

11. a. $0.863 \text{ atm} \times \frac{101.325 \text{ kPa}}{1 \text{ atm}} = 87.4 \text{ kPa}$

   b. $226 \text{ kJ} \times \frac{10^3 \text{ J}}{1 \text{ kJ}} \times \frac{1 \text{ cal}}{4.184 \text{ J}} = 5.40 \times 10^4 \text{ cal}$

13. a. 3 b. 5 c. 4 d. 5 e. 2 f. 1

15. $85.638 \text{ g}/237 \text{ cm}^3 = 0.361 \text{ g/cm}^3$

17. a. $x = 0.240 \text{ g/cm}^3$ b. $x = 12.1 \text{ g/L}$ c. $x = 12.6 \text{ g}$

   d. $x = 0.665 \text{ g/cm}^3$

19. a. 12.27 g b. 32.489 cm c. 75.6 mL d. 126 oz

21. a. $17.5 \text{ qt} \times \frac{1 \text{ L}}{1.057 \text{ qt}} = 16.6 \text{ L}$

   b. $17.5 \text{ qt} \times \frac{1 \text{ L}}{1.057 \text{ qt}} \times \frac{1 \text{ m}^3}{10^3 \text{ L}} = 1.66 \times 10^{-2} \text{ m}^3$

   c. $17.5 \text{ qt} \times \frac{1 \text{ L}}{1.057 \text{ qt}} \times \frac{1 \text{ ft}^3}{28.32 \text{ L}} = 0.585 \text{ ft}^3$

23. $1 \text{ hectare} \times \frac{10^4 \text{ m}^2}{1 \text{ hectare}} \times \frac{(39.37/12 \text{ ft})^2}{1 \text{ m}^2} \times \frac{1 \text{ acre}}{(208.7 \text{ ft})^2} = 2.471 \text{ acre}$

25. a. $1 \text{ Mars day} \times \frac{8.864 \times 10^4 \text{ s}}{1 \text{ Mars day}} \times \frac{1 \text{ h}}{3600 \text{ s}} \times \frac{1 \text{ d}}{24 \text{ h}} = 1.026 \text{ d}$

b. $1 \text{ Mars year} \times \frac{5.935 \times 10^7 \text{ s}}{1 \text{ Mars year}} \times \frac{1 \text{ h}}{3600 \text{ s}} \times \frac{1 \text{ d}}{24 \text{ h}} = 686.9 \text{ d}$

27. a. $6.00 \text{ kides} \times \frac{4 \text{ yards}}{1 \text{ kide}} \times \frac{4 \text{ nookes}}{1 \text{ yard}} \times \frac{2 \text{ fardells}}{1 \text{ nooke}} = 192 \text{ fardells}$

b. $15 \text{ nookes} \times \frac{1 \text{ yard}}{4 \text{ nookes}} \times \frac{1 \text{ kide}}{4 \text{ yards}} = 0.94 \text{ kide}$

$15 \text{ nookes} \times \frac{2 \text{ fardells}}{1 \text{ nooke}} = 30 \text{ fardells}$

29. $\frac{235 \text{ kJ}}{250 \text{ mL}} \times \frac{1 \text{ kcal}}{4.184 \text{ kJ}} \times \frac{1000 \text{ mL}}{1.057 \text{ qt}} \times \frac{1 \text{ qt}}{2 \text{ pt}} \times \frac{1 \text{ pt}}{2 \text{ cup}} = 53.1 \frac{\text{kcal}}{\text{cup}}$

31. $2.507 \text{ g penny} \times \frac{97.6 \text{ g Zn}}{100 \text{ g penny}} = 2.45 \text{ g Zn}$

$2.507 \text{ g penny} \times \frac{2.4 \text{ g Cu}}{100 \text{ g penny}} = 0.060 \text{ g Cu}$

33. a. P b. P c. P d. C

35. d = 0.600 g/0.270 mL = 2.22 g/mL

37. $V_{water} = 18.52 \text{ g} \times \frac{1 \text{ cm}^3}{1.00 \text{ g}} = 18.5 \text{ cm}^3$

$24.5 \text{ cm}^3 - 18.5 \text{ cm}^3 = 6.0 \text{ cm}^3$

$d = 20.32 \text{ g}/6.0 \text{ cm}^3 = 3.4 \text{ g/cm}^3$

39. $(8.0 \times 7.0 \times 0.75) \text{ ft}^3 \times \frac{28.32 \text{ L}}{1 \text{ ft}^3} \times \frac{1 \text{ kg}}{1 \text{ L}} = 1.2 \times 10^3 \text{ kg}$

41. $75 \text{ kg oxygen} \times \frac{1 \text{ L oxygen}}{1.31 \times 10^{-3} \text{ kg}} \times \frac{1 \text{ L}}{0.21 \text{ L oxygen}} = 2.7 \times 10^5 \text{ L}$

43. a. $34.5 \text{ g water} \times \frac{37.0 \text{ g KCl}}{100 \text{ g water}} = 12.8 \text{ g KCl}$

b. $34.5 \text{ g KCl} \times \frac{100 \text{ g water}}{37.0 \text{ g KCl}} = 93.2 \text{ g water}$

45. As; mp = 816°C; d = 5.78 g/mL

47. neutron activation analysis; > 0.0003%

49. a. chemical properties observed during reaction

b. solute is one component of solution
c. compound is pure substance

51. b, c

53. $2.0 \text{ carat} \times \frac{0.200 \text{ g}}{1 \text{ carat}} \times \frac{1 \text{ cm}^3}{3.51 \text{ g}} = 0.11 \text{ cm}^3$

55. $V = 104.2 \text{ g} \times \frac{1 \text{ cm}^3}{4.55 \text{ g}} = 22.9 \text{ cm}^3$

$r = (V/\pi \ell)^{\frac{1}{2}} = (22.9 \text{ cm}^3/\pi \times 4.75 \text{ cm})^{\frac{1}{2}} = 1.24 \text{ cm}$

diameter = 2.48 cm

57. $V_{pyc} = 11.031 \text{ g} \times \frac{1 \text{ cm}^3}{1.000 \text{ g}} = 11.03 \text{ cm}^3$

$V_{water} = 9.994 \text{ g} \times \frac{1 \text{ cm}^3}{1.000 \text{ g}} = 9.994 \text{ cm}^3$

$V_{alloy} = 1.04 \text{ cm}^3 \quad d = 8.240 \text{ g}/1.04 \text{ cm}^3 = 7.92 \text{ g/cm}^3$

58. $t_{°F} = 1.8(2t_{°F}) + 32°$; $-12.3°F = -24.6°C$

59. $V = 63.0 \text{ gal} \times \frac{4 \text{ qt}}{1 \text{ gal}} \times \frac{1 \text{ L}}{1.057 \text{ qt}} \times \frac{1 \text{ m}^3}{10^3 \text{ L}} = 0.238 \text{ m}^3$

$\text{area} = \frac{0.238 \text{ m}^3}{1.2 \times 10^{-7} \text{ m}} = 2.0 \times 10^6 \text{ m}^2 = 2.0 \text{ km}^2$

60. $V = 0.750 \text{ g} \times \frac{1 \text{ cm}^3}{2.70 \text{ g}} = 0.278 \text{ cm}^3$

$r = \frac{0.0179 \text{ in}}{2} \times \frac{2.54 \text{ cm}}{1 \text{ in}} = 0.0227 \text{ cm}$

$\text{length} = \frac{0.278 \text{ cm}^3}{\pi(0.0227 \text{ cm})^2} = 171 \text{ cm}$

61. $\frac{8.50 \times 10^3 \text{ L}}{\text{d}} \times \frac{1 \text{ m}^3}{10^3 \text{L}} \times \frac{7.0 \times 10^{-6} \text{ g Pb}}{1 \text{ m}^3} \times 0.75 \times 0.50 \times \frac{365.2 \text{ d}}{1 \text{ yr}}$

$= 8.1 \times 10^{-3} \text{ g Pb}$

CHAPTER 2
Atoms, Molecules, and Ions

LECTURE NOTES

Students find this material relatively easy to assimilate; it's almost entirely qualitative. On the other hand, there's a lot of memorizing (sorry: learning) involved. This chapter is readily covered in two lectures, perhaps 1½. Some general observations:

- material in Sections 2.1-2.3 is generally well-covered in high school chemistry courses; you need not dwell on it.

- students need to learn the molecular formulas of the elements (Fig. 2.6), charges of transition metal cations (Fig. 2.8), names and formulas of polyatomic ions (Table 2.3) and the nomenclature system for oxoacids.

- naming compounds requires students to distinguish ionic vs molecular substances. It helps to point out that binary molecular compounds are composed of two nonmetals. Almost all ionic compounds contain a metal cation.

- the periodic table will be discussed in greater detail later in the text.

<u>LECTURE 1</u>

I <u>Atoms</u>

A. <u>Atomic theory</u>
Postulates: Elements consist of tiny particles called atoms, which retain their identity in reactions. In a compound, atoms of two or more elements are combined in a fixed ratio of small whole numbers (e.g., 1:1, 2:1, 3:2, etc.).

B. <u>Components</u>

	relative mass	relative charge	location
proton	1	+1	nucleus
neutron	1	0	nucleus
electron	0.0005	-1	outside

C. <u>Atomic number</u> = number of protons in nucleus = number of electrons in neutral atom. Characteristic of a particular element; all H atoms have 1 proton, all He atoms have 2 protons, etc.

D. Mass number = number of protons + number of neutrons. Atoms of same element can differ in mass number

	prot.	neut.	at.no.	mass no.	nucl. symbol
carbon-12	6	6	6	12	$^{12}_{6}C$
carbon-14	6	8	6	14	$^{14}_{6}C$

$^{12}_{6}C$ and $^{14}_{6}C$ are referred to as isotopes

II Periodic Table

A. Structure Periods and groups; numbering system for groups

B. Metals located at lower left of table, nonmetals at upper right. Metalloids

III Molecules

A. Usually made up of nonmetal atoms; held together by covalent bonds.

B. Types of formulas

Consider the compound ethane

Molecular formula: C_2H_6 Simplest formula: CH_3

Structural formula:

```
     H   H
     |   |
 H - C - C - H
     |   |
     H   H
```

LECTURE 2

I Ions

A. Formation of monatomic ions

$$\text{Na atom } (11\ p^+, 11\ e^-) \longrightarrow \text{Na}^+ \text{ ion } (11\ p^+, 10\ e^-) + e^-$$

$$\text{F atom } (9\ p^+, 9\ e^-) + e^- \longrightarrow \text{F}^- \text{ ion } (9\ p^+, 10\ e^-)$$

Nucleus remains unchanged

B. Charges of monatomic ions

1. Ions with noble gas structures

Group 1 (+1); Group 2 (+2); Al^{3+}

Group 16 (-2); Group 17 (-1); N^{3-}

2. Transition metal cations (Figure 2.8)

C. Polyatomic ions

General structure. Names and formulas (Table 2.3)

D. Formulas Apply principle of electrical neutrality

calcium fluoride: Ca^{2+}, F^- ions: CaF_2

aluminum nitrate: Al^{3+}, NO_3^- ; $Al(NO_3)_3$

sodium dihydrogen phosphate: Na^+, $H_2PO_4^-$; NaH_2PO_4

II Names of Compounds

A. Ionic Name cation, followed by anion. Note that with transition metal cations, charge is indicated by Roman numeral.

Na_2SO_4 sodium sulfate $Fe(NO_3)_3$ iron(III) nitrate

NH_4Br ammonium bromide

Note that, except for ammonium salts, ionic compounds contain metal cations.

Systematic names of oxoanions (ate, ite; per, hypo)

calcium hypochlorite: $Ca(ClO)_2$

B. Binary molecular compounds Use of Greek prefixes

SF_6 sulfur hexafluoride N_2O_3 dinitrogen trioxide

C. Acids

binary acids: hydrochloric acid

oxoacids: ate salt $\rightarrow$ ic acid

$HClO_4$ perchloric acid $Ca(ClO_4)_2$ calcium perchlorate

DEMONSTRATIONS

1. Conservation of mass: Test. Dem. 54
2. Reaction of iron with chlorine: Shak. 1 66
3. Hydrates of cobalt(II) chloride: J. Chem. Educ. 68 779 (1991)

QUIZZES

Quiz 1

1. Consider the element cobalt (Z = 27)
   a. Classify Co as a metal, nonmetal, or metalloid
   b. State the number of protons, neutrons and electrons in ${}^{55}_{27}Co^{2+}$
2. Write the formulas of
   a. ammonium sulfate   b. potassium carbonate
   c. phosphorus pentachloride   d. calcium chlorite
3. Name the following compounds
   a. $Al_2(SO_4)_3$ b. $FeBr_3$ c. N_2O_4

Quiz 2

1. Give the simplest formula of
   a. C_3H_8 b. C_6H_6 c. N_2H_4
2. Name:
   a. KNO_3 b. $CoSO_4$ c. $HClO_4$ d. P_4O_6
3. Give the formula of
   a. chlorine gas   b. aluminum bromide   c. carbon tetrachloride
   d. sodium carbonate

Quiz 3

1. Give the nuclear symbol of a species containing 29 protons, 27 electrons, and 31 neutrons. What would be the formula of the compound between this cation and the sulfate anion?
2. Name the acid derived from each of the following compounds by replacing metal cations with hydrogen atoms.
   a. K_2SO_4 b. KClO   c. KNO_3 d. KNO_2
3. Give the formulas of all the compounds containing no ions other than the following:

   Al^{3+}, NH_4^+, SO_4^{2-}, Cl^-

Quiz 4

1. Arrange in order of increasing mass: p^+, e^-, ${}^{1}_{1}H$, ${}^{2}_{1}H$
2. Give the formulas of the following ionic compounds:
   a. lead(II) nitrate   b. chromium(III) sulfate
   c. magnesium phosphate
3. Name:
   a. Cu_2SO_4 b. P_4O_6 c. $MnCO_3$ d. $Fe(ClO_4)_3$

Quiz 5

1. Give the number of protons in
   a. Co^{2+} b. NH_4^+ c. 3 Al atoms   d. OH^-
2. What is the name of P_4O_{10}? What is its simplest formula? Are the elements in this compound metals, nonmetals, or metalloids? Does this formula represent a molecular or an ionic compound?
3. Write formulas for
   a. magnesium hydrogen carbonate   b. chloric acid
   c. ammonium sulfide   d. cobalt(III) sulfate

Answers

Quiz 1: 1. a. metal b. 27, 28, 25

2. a. $(NH_4)_2SO_4$ b. K_2CO_3 c. PCl_5 d. $Ca(ClO_2)_2$

3. a. aluminum sulfate b. iron(III) bromide
   c. dinitrogen tetroxide

Quiz 2: 1. a. C_3H_8 b. CH c. NH_2

2. a. potassium nitrate b. cobalt(II) sulfate
   c. perchloric acid d. tetraphosphorus hexaoxide

3. a. Cl_2 b. $AlBr_3$ c. CCl_4 d. Na_2CO_3

Quiz 3: 1. $^{60}_{29}Cu^{2+}$; $CuSO_4$

2. a. sulfuric acid b. hypochlorous acid
   c. nitric acid d. nitrous acid

3. $Al_2(SO_4)_3$, $AlCl_3$, $(NH_4)_2SO_4$, NH_4Cl

Quiz 4: 1. $e^- < p^+ < {}^1_1H < {}^2_1H$

2. a. $Pb(NO_3)_2$ b. $Cr_2(SO_4)_3$ c. $Mg_3(PO_4)_2$

3. a. copper(I) sulfate b. tetraphosphorus hexaoxide
   c. manganese(II) carbonate d. iron(III) perchlorate

Quiz 5: 1. a. 27 b. 11 c. 39 d. 9

2. tetraphosphorus decaoxide; P_2O_5; nonmetals; molecular

3. a. $Mg(HCO_3)_2$ b. $HClO_3$ c. $(NH_4)_2S$ d. $Co_2(SO_4)_3$

PROBLEMS

1. a. NH_3 b. H_2O c. CH_4 d. HCl

3. a. $CuSO_4(s)$; white b. $CoCl_2\ 6H_2O(s)$; pink

5. Dalton; compound always contains same elements in same mass ratio

7. a. conservation of mass b. constant composition
   c. multiple proportions d. none

9. 1st experiment: 3.56 g Mg, 2.37 g O (60.0% Mg)
   2nd experiment: 1.65 g Mg, 1.10 g O (60.0% Mg)

11. J. J. Thompson; cathode ray studies (Section 2.2)

13. 34 protons, 46 neutrons

15. a. no; in second case, no. of neutrons not specified
    b. yes; atomic number of Fe must be 26

17. a. 3 b. 4 c. 3 d. 4, 3, 2

19.

F	___	___	___	9
___	___	15	___	15
Fe	___	___	___	23
S	-2	___	___	___

21. a. 16 p^+, 18 e^- b. 128 p^+, 128 e^- c. 18 p^+, 18 e^-
d. 1 p^+, 0 e^-

23.

___	30	25	23
${}^{80}_{34}Se^{2-}$	___	___	___
___	116	78	75

25. a. K b. Cd c. Au d. Sb e. Rb

27. All metals except Sb, which is a metalloid

29. a. 3 b. 5 c. 6 d. 4

31.

ammonia	___	NH_3
___	Se_2Cl_2	SeCl
xenon trioxide	___	XeO_3
bromine trichloride	___	$BrCl_3$
diphosphorus pentoxide	___	P_2O_5
___	N_2O_4	NO_2

33. KBr, K_2O, $SrBr_2$, SrO

35. a. $Fe_2(SO_4)_3$ b. $KC_2H_3O_2$ c. PbO d. $Ba(ClO_3)_2$
e. $CaSO_4$

37. a. $Na_2Cr_2O_7$ b. $AlPO_4$ c. Cr_2O_3 d. $Ca_3(PO_4)_2$
e. H_2SO_3

39. Ba_3N_2, $Cu(OH)_2$, silver(I) telluride, iron(III) carbonate, $SrCrO_4$, sodium hydrogen carbonate, NH_4ClO_4

41. a. perchloric acid   b. chlorous acid   c. hypoiodous acid

d. nitric acid

43. a. $KClO_2$ b. $Ca(NO_2)_2$ c. Na_2SO_3 d. $NaClO$

45. a. $Na_2CO_3 \cdot 10H_2O$ b. $MgSO_4 \cdot 7H_2O$ c. $Na_2SO_4 \cdot 10H_2O$

d. $CoCl_2 \cdot 6H_2O$

47. Heat to drive off water

49. a. depends on charges of ions
b. ionic compounds do not have "molecular formulas"
c. only for a few light atoms
d. only in a cation

51. a. iodine   b. chlorine   c. rhodium   d. tellurium

53. a. ethane: 18.0 g C/4.53 g H = 3.97 g C/g H

ethene: 43.20 g C/7.25 g H = 5.96 g C/g H

5.96/3.97 = 1.50 = 3/2

b. CH_2 and CH_3; C_2H_4 and C_2H_6; - - -

54. mass = $13(1.6727 \times 10^{-24}\text{ g}) + 13(9.1095 \times 10^{-28}\text{ g})$

$+ 14(1.6750 \times 10^{-24}\text{ g}) = 4.5207 \times 10^{-23}\text{ g}$

$V = \frac{4\pi}{3}(1.43 \times 10^{-8}\text{ cm})^3 = 1.22 \times 10^{-23}\text{ cm}^3$

$d = 4.5207\text{ g}/1.22\text{ cm}^3 = 3.71\text{ g/cm}^3$

Empty space between Al atoms

55. $1.4965 \times 10^{-23}\text{ g} - 2(9.1095 \times 10^{-28}\text{ g}) = 1.4963 \times 10^{-23}\text{ g}$

56. a. $200 \times 500 \times 2.5 \times 10^{19}$ molecules = 2.5×10^{24}

b. $(2.5 \times 10^{24})/(1.1 \times 10^{44}) = 2.3 \times 10^{-20}$

c. $(2.3 \times 10^{-20}) \times (500) \times (2.5 \times 10^{19}) \approx 2.9 \times 10^2$

CHAPTER 3
Mass Relations in Chemistry; Stoichiometry

LECTURE NOTES

This chapter is considerably more difficult and time-consuming than the two that precede it. It contains a considerable amount of quantitative material that is fundamental for future chapters. We suggest you devote three lectures to Chapter 3. The first lecture deals with atomic masses and the mole (Sections 3.1, 3.2), the second with the quantitative aspects of chemical formulas (Section 3.3), the third with mass relations in reactions (Section 3.4). Points to keep in mind include the following:

1. Students have little trouble calculating atomic masses from isotopic data. The reverse process, estimating isotopic abundances from atomic masses, is much more difficult because it involves solving an algebraic equation. Incidentally, the treatment of significant figures here is a bit tricky because the percent abundances are dependent on one another. In general, the atomic mass can be calculated with more precision than you might suppose at first glance from the abundance data.

2. Formula mass is introduced largely to make the calculation of masses of molecules a little easier to follow (Example 3.3); it could be deleted.

3. Note that mole-gram conversions (Section 3.2) will be required in many later chapters, often as the first step in a more complex problem.

4. When dealing with formulas (Section 3.3), it is important to emphasize early on that the subscripts give not only the atom ratio but also the mole ratio. It is necessary that students realize this if they are to follow the logic of obtaining simplest formulas from mass percents.

5. Students ordinarily have little trouble calculating formulas from mass percents. They are much less adept at obtaining formulas from analytical data such as that given in Example 3.9.

6. It is important to get across the point (Section 3.4) that a chemical equation describes what happens when a reaction is carried out in the laboratory. Including in the equation

the physical states of reactants and products helps to emphasize this point.

7. When you discuss mass relations in reactions, some students will revert to the infamous "ratio-and-proportion" method. The comments made in Chapter 1 about conversion factors apply here too.

8. There are many different ways to find the limiting reactant and calculate the theoretical yield. We've tried most of them and recommend the approach described in Section 3.4.

LECTURE 1

I Atomic and Formula Masses

A. Meaning of atomic masses - give relative masses of atoms. Based on C-12 scale; most common isotope of carbon is assigned an atomic mass of exactly 12 amu.

element	B	Ca	Ni
atomic mass	10.81 amu	40.08 amu	58.69 amu

A nickel atom is 58.69/40.08 = 1.464 times as heavy as a calcium atom. It is 58.69/10.81 = 5.429 times as heavy as a boron atom.

B. Atomic masses from isotopic composition
A. M. = (A.M. isotope 1)(%/100) + (A.M. isotope 2)(%/100) +

Isotope	Atomic mass	Percent
Ne-20	20.00 amu	90.92
Ne-21	21.00 amu	0.26
Ne-22	22.00 amu	8.82

A. M. Ne = 20.00(0.9092) + 21.00(0.0026) + 22.00(0.0882)
= 20.18 amu

C. Masses of individual atoms Since the atomic masses of H, Cl and Ni are 1.008 amu, 35.45 amu, and 58.69 amu, it follows that:

1.008 g H, 35.45 g Cl, 58.69 g Ni

all contain the same number of atoms, N_A. It turns out that

$$N_A = \text{Avogadro's number} = 6.022 \times 10^{23}$$

1. Mass of H atom:

$$1 \text{ atom H} \times \frac{1.008 \text{ g H}}{6.022 \times 10^{23} \text{ atoms}} = 1.674 \times 10^{-24} \text{ g}$$

2. Number of atoms in one gram of nickel:

$$1.000 \text{ g Ni} \times \frac{6.022 \times 10^{23} \text{ atoms Ni}}{58.69 \text{ g Ni}} = 1.026 \times 10^{22} \text{ atoms}$$

D. Formula mass = sum of atomic masses in formula

FM H_2O = 18.02 amu

FM NaCl = 58.44 amu

II The Mole

A. Meaning 1 mol = 6.022×10^{23} items

1 mol H = 6.022×10^{23} H atoms; mass = 1.008 g

1 mol Cl = 6.022×10^{23} Cl atoms; mass = 35.45 g

1 mol Cl_2 = 6.022×10^{23} Cl_2 molecules; mass = 70.90 g

1 mol HCl = 6.022×10^{23} HCl molecules; mass = 36.46 g

B. Molar mass Generalizing from the above examples, the molar mass, $\mathcal{M}$, is numerically equal to the formula mass.

	formula mass	molar mass
$CaCl_2$	110.98 amu	110.98 g/mol
$C_6H_{12}O_6$	180.18 amu	180.18 g/mol

LECTURE 2

I The Mole

A. Mole-mass conversions

1. Calculate mass in grams of 13.2 mol of $CaCl_2$

$$\text{mass} = 13.2 \text{ mol } CaCl_2 \times \frac{110.98 \text{ g } CaCl_2}{1 \text{ mol } CaCl_2} = 1.47 \times 10^3 \text{ g}$$

2. Calculate number of moles in 16.4 g of $C_6H_{12}O_6$

$$\text{no. moles} = 16.4 \text{ g } C_6H_{12}O_6 \times \frac{1 \text{ mol } C_6H_{12}O_6}{180.18 \text{ g } C_6H_{12}O_6} = 0.0910 \text{ mol}$$

II Formulas

A. Mass percent from formula

Percent composition of K_2CrO_4?

molar mass = (78.20 + 52.00 + 64.00)g/mol = 194.20 g/mol

$$\% \text{ K} = \frac{78.20}{194.20} \times 100 = 40.27 \qquad \% \text{ Cr} = \frac{52.00}{194.20} \times 100 = 26.78$$

$$\% \text{ O} = \frac{64.00}{194.20} \times 100 = 32.96$$

Note that percents must add to 100

B. Simplest formula from percent composition

Find mass of each element in sample of compound. Then find numbers of moles of each element and finally the mole ratio.

Simplest formula of compound containing 26.6% K, 35.4% Cr and 38.0% O?

Work with 100 g sample: 26.6 g K, 35.4 g Cr, 38.0 g O

$$\text{no. moles K} = 26.6 \text{ g} \times \frac{1 \text{ mol}}{39.10 \text{ g}} = 0.680 \text{ mol K}$$

$$\text{no. moles Cr} = 35.4 \text{ g} \times \frac{1 \text{ mol}}{52.00 \text{ g}} = 0.681 \text{ mol Cr}$$

$$\text{no. moles O} = 38.0 \text{ g} \times \frac{1 \text{ mol}}{16.00 \text{ g}} = 2.38 \text{ mol O}$$

Note that 2.38/0.680 = 3.50 = 7/2 Simplest formula: $K_2Cr_2O_7$

C. Simplest formula from analytical data

A sample of acetic acid (C, H, O atoms) weighing 1.000 g burns to give 1.466 g CO_2 and 0.6001 g H_2O. Simplest formula?

Find mass of C in sample (from CO_2), then mass of H (from H_2O), and finally mass of O by difference.

$$\text{mass C} = 1.466 \text{ g } CO_2 \times \frac{12.01 \text{ g C}}{44.01 \text{ g } CO_2} = 0.4001 \text{ g C}$$

$$\text{mass H} = 0.6001 \text{ g } H_2O \times \frac{2.02 \text{ g H}}{18.02 \text{ g } H_2O} = 0.0673 \text{ g H}$$

mass O = 1.000 g - 0.400 g - 0.067 g = 0.533 g

$$\text{no. moles C} = 0.4001 \text{ g C} \times \frac{1 \text{ mol C}}{12.01 \text{ g C}} = 0.0333 \text{ mol C}$$

$$\text{no. moles H} = 0.0673 \text{ g H} \times \frac{1 \text{ mol H}}{1.008 \text{ g H}} = 0.0666 \text{ mol H}$$

$$\text{no. moles O} = 0.533 \text{ g O} \times \frac{1 \text{ mol O}}{16.00 \text{ g O}} = 0.0333 \text{ mol O}$$

Simplest formula is CH_2O

D. Molecular formula from simplest formula

Must know molar mass. For acetic acid, $\mathcal{M}$ = 60 g/mol
Formula mass = 30 amu; 60/30 = 2
Molecular formula: $C_2H_4O_2$

LECTURE 3

I Chemical Equations

A. Balancing Must have same number of atoms of each type on both sides. Achieve this by adjusting coefficients in front of formulas. Example: combustion of propane in air to give carbon dioxide and water.

$$C_3H_8(g) + O_2(g) \longrightarrow CO_2(g) + H_2O(l)$$

Balance C: $C_3H_8(g) + O_2(g) \longrightarrow 3CO_2(g) + H_2O(l)$

Balance H: $C_3H_8(g) + O_2(g) \longrightarrow 3CO_2(g) + 4H_2O(l)$

Balance O: $C_3H_8(g) + 5O_2(g) \longrightarrow 3CO_2(g) + 4H_2O(l)$

Meaning: 1 mol C_3H_8 reacts with 5 mol O_2 to form 3 mol CO_2 and 4 mol H_2O

B. Mass relations in reactions

1. Moles of CO_2 produced when 1.65 mol C_3H_8 burns?

   Use coefficients of balanced equation to obtain conversion factor.

   $$1.65 \text{ mol } C_3H_8 \times \frac{3 \text{ mol } CO_2}{1 \text{ mol } C_3H_8} = 4.95 \text{ mol } CO_2$$

2. Mass of O_2 required to react with 12.0 g of C_3H_8?

   $$12.0 \text{ g } C_3H_8 \times \frac{1 \text{ mol } C_3H_8}{44.09 \text{ g } C_3H_8} \times \frac{5 \text{ mol } O_2}{1 \text{ mol } C_3H_8} \times \frac{32.00 \text{ g } O_2}{1 \text{ mol } O_2}$$

   $= 43.6$ g O_2

II Yield of Product in Reaction

A. Limiting reactant, theoretical yield. Ordinarily, reactants are not present in the exact ratio required for reaction. Instead, one reactant is in excess; some of it is left when the reaction is over. The other, limiting reactant, is completely consumed to give the theoretical yield of product.

To calculate the theoretical yield and identify the limiting reactant:

1. Calculate the yield expected if the first reactant is limiting.
2. Repeat this calculation for the second reactant.
3. The theoretical yield is the smaller of these two quantities. The reactant that gives the smaller calculated yield is the limiting reactant.

$$2Ag(s) + I_2(s) \longrightarrow 2AgI(s)$$

Calculate the theoretical yield of AgI and determine the limiting reactant starting with 1.00 g of Ag and 1.00 g of I_2.

theor. yield AgI if Ag is limiting:

$$1.00 \text{ g Ag} \times \frac{469.54 \text{ g AgI}}{215.74 \text{ f Ag}} = 2.18 \text{ g AgI}$$

theor. yield if I_2 is limiting:

$$1.00 \text{ g } I_2 \times \frac{469.54 \text{ g AgI}}{253.80 \text{ g } I_2} = 1.85 \text{ g AgI}$$

Theoretical yield = 1.85 g AgI; I_2 is limiting

B. Actual yield, percent yield

$$\% \text{ yield} = \frac{\text{Actual Yield}}{\text{Theoretical Yield}} \times 100$$

Suppose actual yield of AgI were 1.50 g:

$$\% \text{ yield} = \frac{1.50}{1.85} \times 100 = 81.1$$

DEMONSTRATIONS

1. Combustion of propane: J. Chem. Educ. 64 894 (1987)
2. Reaction of antimony with halogens: Test. Dem. 43; Shak. 1 64

QUIZZES

Quiz 1

1. Calculate the atomic mass of gallium, which consists of two isotopes, Ga-69 (mass = 68.92 amu, abundance = 60.16%) and Ga-71 (mass = 70.92 amu, abundance = 39.84%).
2. Determine the simplest formula of a compound of carbon, hydrogen and chlorine in which the mass percent of C is 49.02 and that of H is 2.74.

Quiz 2

1. Determine the mass in grams and the number of atoms in 2.31 mol of Fe.
2. Consider the reaction: $2Cr(s) + 3Cl_2(g) \longrightarrow 2CrCl_3(s)$

   a. What mass of $CrCl_3$ can be formed from 62.5 g of Cl_2?

b. What is the theoretical yield of $CrCl_3$, starting with 1.00 g of Cr and 2.00 g of Cl_2?

Quiz 3

1. Calculate the mass in grams of 0.0138 mol of PCl_3.
2. Combustion of a 1.000 g sample of a compound containing the three elements C, H, and Cl gives 2.325 g of carbon dioxide and 0.397 g of water. What are the mass percents of C, H, and Cl in the compound?

Quiz 4

1. Chlorine (A.M. = 35.45 amu) consists of Cl-35 (A.M. = 34.98 amu) and Cl-37 (A.M. = 36.98 amu). Estimate the abundances of these two isotopes.
2. Consider the reaction:

$$Fe^{3+}(aq) + 3\ OH^-(aq) \longrightarrow Fe(OH)_3(s)$$

What is the theoretical yield of product, starting with

a. 1.20 mol Fe^{3+}, 2.62 mol OH^-?

b. 1.00 g Fe^{3+}, 1.00 g OH^-?

Quiz 5

1. How many CH_4 molecules are required to weigh one gram?
2. Consider the reaction between silver and sulfur.
   a. Write a balanced equation for the reaction; the product is silver(I) sulfide.
   b. Calculate the number of moles of silver required to react with 15.0 g of sulfur.
   c. How many grams of sulfur are required to react with 1.00 g of silver?

Answers

Quiz 1 1. 69.72 amu 2. C_3H_2Cl

Quiz 2 1. 129 g, 1.39×10^{24} atoms 2. a. 93.1 g b. 2.98 g

Quiz 3 1. 1.90 g 2. 63.45% C, 4.44% H, 32.11% Cl

Quiz 4 1. 24% Cl-37, 76% Cl-35 2. a. 0.873 mol b. 1.91 g

Quiz 5 1. 3.754×10^{22} atoms

2. a. $2Ag(s) + S(s) \longrightarrow Ag_2S(s)$

b. 0.936 mol Ag c. 6.73 g

PROBLEMS

1. a. 45.98 amu b. 72.90 amu c. 39.10 amu d. < 39.10 amu

   d < c < a < b

3. 83.9134 amu(0.005) + 85.9094 amu(0.099) + 86.9089 amu(0.070) + 87.9056 amu(0.826) = 87.62 amu

5. 10.811 amu = 10.013 amu(x) + 11.089 amu(1 - x); x = 0.199

   19.9% B-10, 80.1% B-11

7. 24.305 = 23.98(0.899 - x) + 24.98(0.101) + 25.98 x; x = 0.11

   79% Mg-24, 11% Mg-26

9. Major peak at mass 28, small peak at mass 29, still smaller peak at mass 30.

11. a. $1 \text{ W atom} \times \dfrac{183.9 \text{ g}}{6.022 \times 10^{23} \text{ W atoms}} = 3.054 \times 10^{-22} \text{ g}$

    b. $1.000 \times 10^{-3} \text{ g W} \times \dfrac{6.022 \times 10^{23} \text{ W atoms}}{183.9 \text{ g}} = 3.275 \times 10^{18} \text{ atoms}$

13. a. $10^{9} \text{ Au atoms} \times \dfrac{197.0 \text{ g}}{6.022 \times 10^{23} \text{ atoms}} = 3.271 \times 10^{-13} \text{ g}$

    b. $1 \text{ oz Au} \times \dfrac{453.6 \text{ g}}{16 \text{ oz}} \times \dfrac{6.022 \times 10^{23} \text{ atoms}}{197.0 \text{ g}} = 8.666 \times 10^{22} \text{ atoms}$

15. $3.2333 \times 10^{14} \text{ pennies} \times \dfrac{1 \text{ mol}}{6.022 \times 10^{23} \text{ pennies}} = 5.369 \times 10^{-10} \text{ mol}$

17. a. 13 e^-

    b. $6.022 \times 10^{23} \text{ atoms} \times 13 \text{ e}^-/\text{atom} = 7.829 \times 10^{24} \text{ e}^-$

    c. $0.2843 \text{ mol} \times \dfrac{7.829 \times 10^{24} \text{ e}^-}{1 \text{ mol}} = 2.226 \times 10^{24} \text{ e}^-$

    d. $0.2843 \text{ g} \times \dfrac{1 \text{ mol}}{26.98 \text{ g}} \times \dfrac{7.829 \times 10^{24} \text{ e}^-}{1 \text{ mol}} = 8.250 \times 10^{22} \text{ e}^-$

19. a. 69.72 g/mol

    b. [2(14.01) + 16.00] g/mol = 44.02 g/mol

    c. [12(12.01) + 22(1.008) + 11(16.00)] g/mol = 342.30 g/mol

21. a. $\mathcal{M}$ = 334.41 g/mol

    $0.830 \text{ g} \times \dfrac{1 \text{ mol}}{334.41 \text{ g}} = 2.48 \times 10^{-3} \text{ mol}$

b. $\mathcal{M}$ = 180.15 g/mol

$$0.25000 \text{ g} \times \frac{1 \text{ mol}}{180.15 \text{ g}} = 1.3877 \times 10^{-3} \text{ mol}$$

c. $\mathcal{M}$ = 176.12 g/mol

$$1.0000 \text{ g} \times \frac{1 \text{ mol}}{176.12 \text{ g}} = 5.6779 \times 10^{-3} \text{ mol}$$

23. a. 5.75 mol x 14.01 g/mol = 80.6 g
    b. 5.75 mol x 28.02 g/mol = 161 g
    c. 5.75 mol x 17.03 g/mol = 97.9 g

25. $\mathcal{M}$ = 62.07 g/mol

    a. 0.1245 g, 2.006×10^{-3} mol, 1.208×10^{21} molecules, 2.416×10^{21} C atoms

    b. 2.33 g, 0.0375 mol, 2.26×10^{22} molecules, 4.52×10^{22} C atoms

    c. 2.0×10^{3} g, 33 mol, 2.0×10^{25} molecules, 4.0×10^{25} C atom

    d. 1.9×10^{-10} g, 3.0×10^{-12} mol, 1.8×10^{12} molecules, 3.6×10^{12} C atoms

27. $\mathcal{M}$ = 813.44 g/mol

$$\% \text{ Cu} = \frac{63.55}{813.44} \times 100 = 7.812 \qquad \% \text{ Al} = \frac{161.88}{813.44} \times 100 = 19.901$$

$$\% \text{ P} = \frac{123.88}{813.44} \times 100 = 15.229 \qquad \% \text{ O} = \frac{448.00}{813.44} \times 100 = 55.075$$

$$\% \text{ H} = \frac{16.128}{813.44} \times 100 = 1.9827$$

29. $\mathcal{M}$ = 776.8 g/mol

$$5.00 \text{ g Thy} \times \frac{507.6 \text{ g I}}{776.8 \text{ g Thy}} \times \frac{10^{3} \text{ mg}}{1 \text{ g}} = 3.27 \times 10^{3} \text{ mg}$$

31. a. $\% \text{ aspirin} = \frac{152}{250.0} \times 100 = 60.8$

    b. $\mathcal{M}$ = 180.15 g/mol

$$0.611 \text{ g} \times 0.608 \times \frac{108.09}{180.15} = 0.223 \text{ g C}$$

33. $3.348 \text{ g } CO_2 \times \frac{12.01 \text{ g C}}{44.01 \text{ g } CO_2} = 0.9136 \text{ g C}$

91.36% C, 8.64 % H

35. $n \text{ As} = 1.587 \text{ g As} \times \frac{1 \text{ mol As}}{74.92 \text{ g As}} = 0.02118 \text{ mol As}$

$n \text{ Cl} = 3.755 \text{ g Cl} \times \frac{1 \text{ mol Cl}}{35.45 \text{ g Cl}} = 0.1059 \text{ mol Cl}$

0.1059 mol Cl/0.02118 mol As = 5.000 mol Cl/mol As $AsCl_5$

37. a. In 100 g of citric acid, there are 3.123 mol C, 4.17 mol H, 3.643 mol O

4.17 mol H/3.123 mol C = 4/3 mol H/mol C

3.643 mol O/3.123 mol C = 7/6 mol O/mol C $C_6H_8O_7$

b. 2.474 mol C, 6.185 mol H, 0.3092 mol Pb

2.474 mol C/0.3092 mol Pb = 8 mol C/mol Pb

6.185 mol H/0.3092 mol Pb = 20 mol H/mol Pb PbC_8H_{20}

c. 3.822 mol C, 2.73 mol H, 1.638 mol O, 0.5457 mol S, 0.546 mol N

7 mol C/mol N; 5 mol H/mol N; 3 mol O/mol N; 1 mol S/mol N

$C_7H_5O_3SN$

39. mass C = 1.58 g C; mass H = 0.132 g H; mass O = 0.79 g O

n C = 0.132 mol; n H = 0.131 mol; n O = 0.049 mol

8/3 mol C/mol O; 8/3 mol H/mol O

$C_8H_8O_3$

41. 1.470 g C, 0.1707 g H, 0.2142 g N, 0.732 g O

0.1224 mol C, 0.1693 mol H, 0.01529 mol N, 0.0458 mol O

8 mol C/mol N, 11 mol H/mol N, 3 mol O/mol N

$C_8H_{11}O_3N$

43. 1.143 g C, 0.3835 g H, 1.332 g N

0.09517 mol C, 0.3805 mol H, 0.09507 mol N

4 mol H/mol N, 1 mol C/mol N

Simplest formula: CH_4N $\mathcal{M}$ = 30.05 g/mol

Molecular formula: $C_2H_8N_2$

45. $n\ Cl = 0.3059\ g\ HCl \times \frac{35.45\ g\ Cl}{36.46\ g\ HCl} \times \frac{1\ mol\ Cl}{35.45\ g\ Cl}$

$= 0.008390\ mol\ Cl$

$n\ O = 0.5287\ g\ H_2O \times \frac{16.00\ g\ O}{18.02\ g\ H_2O} \times \frac{1\ mol\ O}{16.00\ g\ O} = 0.02934\ mol\ O$

7/2 mol O/mol Cl $\quad Cl_2O_7$

47. $n\ Na_2CO_3 = 1.006\ g \times \frac{1\ mol}{105.99\ g} = 9.492 \times 10^{-3}\ mol\ Na_2CO_3$

$n\ H_2O = 1.708\ g \times \frac{1\ mol}{18.02\ g} = 9.478 \times 10^{-2}\ mol\ H_2O$

$Na_2CO_3 \cdot 10H_2O$

49. a. $Au_2S_3(s) + 3H_2(g) \longrightarrow 2Au(s) + 3H_2S(g)$

b. $C_3H_8(g) + 5\ O_2(g) \longrightarrow 3CO_2(g) + 4H_2O(g)$

c. $SiO_2(s) + 2C(s) \longrightarrow Si(s) + 2CO(g)$

51. a. Na^+, N^{3-} ions: $6Na(s) + N_2(g) \longrightarrow 2Na_3N(s)$

b. Na^+, O^{2-} ions: $4Na(s) + O_2(g) \longrightarrow 2Na_2O(s)$

c. Na^+, S^{2-} ions: $2Na(s) + S(s) \longrightarrow Na_2S(s)$

d. Na^+, Br^- ions: $2Na(s) + Br_2(l) \longrightarrow 2NaBr(s)$

e. Na^+, I^- ions: $2Na(s) + I_2(s) \longrightarrow 2NaI(s)$

53. a. $BF_3(g) + 3H_2O(l) \longrightarrow 3HF(l) + H_3BO_3(s)$

b. $3MgO(s) + 2Fe(s) \longrightarrow Fe_2O_3(s) + 3Mg(s)$

c. $2N_2O(g) \longrightarrow 2N_2(g) + O_2(g)$

d. $CaC_2(s) + 2H_2O(l) \longrightarrow C_2H_2(g) + Ca(OH)_2(s)$

e. $CaCN_2(s) + 3H_2O(l) \longrightarrow CaCO_3(s) + 2NH_3(g)$

55. a. $12.3\ mol\ NO \times \frac{4\ mol\ NH_3}{6\ mol\ NO} = 8.20\ mol\ NH_3$

b. $5.87\ mol\ NO \times \frac{5\ mol\ N_2}{6\ mol\ NO} = 4.89\ mol\ N_2$

c. $0.2384\ mol\ N_2 \times \frac{6\ mol\ NO}{5\ mol\ N_2} = 0.2861\ mol\ NO$

d. $13.9 \text{ mol } NH_3 \times \frac{6 \text{ mol } H_2O}{4 \text{ mol } NH_3} = 20.8 \text{ mol } H_2O$

57. a. $2.93 \text{ mol NO} \times \frac{5 \text{ mol } N_2}{6 \text{ mol NO}} \times \frac{28.02 \text{ g } N_2}{1 \text{ mol } N_2} = 68.4 \text{ g } N_2$

b. $7.65 \text{ mol } H_2O \times \frac{4 \text{ mol } NH_3}{6 \text{ mol } H_2O} \times \frac{17.03 \text{ g } NH_3}{1 \text{ mol } NH_3} = 86.9 \text{ g } NH_3$

c. $0.356 \text{ g } H_2O \times \frac{1 \text{ mol } H_2O}{18.02 \text{ g } H_2O} \times \frac{6 \text{ mol NO}}{6 \text{ mol } H_2O} \times \frac{30.01 \text{ g NO}}{1 \text{ mol NO}} = 0.593 \text{ g NO}$

d. $20.0 \text{ g NO} \times \frac{1 \text{ mol NO}}{30.01 \text{ g NO}} \times \frac{4 \text{ mol } NH_3}{6 \text{ mol NO}} \times \frac{17.03 \text{ g } NH_3}{1 \text{ mol } NH_3}$

$= 7.57 \text{ g } NH_3$

59. a. $Fe_2O_3(s) + 3C(s) \longrightarrow 2Fe(s) + 3CO(g)$

b. $12.79 \text{ mol Fe} \times \frac{1 \text{ mol } Fe_2O_3}{2 \text{ mol Fe}} = 6.395 \text{ mol } Fe_2O_3$

c. $13.68 \text{ g C} \times \frac{1 \text{ mol C}}{12.01 \text{ g C}} \times \frac{3 \text{ mol CO}}{3 \text{ mol C}} \times \frac{28.01 \text{ g CO}}{1 \text{ mol CO}} = 31.90 \text{ g CO}$

61. a. $(2.85 \times 10^3 \times 0.045)\text{g alcohol} \times \frac{1 \text{ mol alcohol}}{46.07 \text{ g alcohol}}$

$\times \frac{1 \text{ mol glucose}}{2 \text{ mol alcohol}} \times \frac{180.16 \text{ g glucose}}{1 \text{ mol glucose}} = 2.5 \times 10^2 \text{ g glucose}$

b. $(2.85 \times 10^3 \times 0.045)\text{g alcohol} \times \frac{1 \text{ mol alcohol}}{46.07 \text{ g alcohol}} \times$

$\frac{2 \text{ mol } CO_2}{2 \text{ mol alcohol}} \times \frac{44.01 \text{ g } CO_2}{1 \text{ mol } CO_2} \times \frac{1 \text{ L}}{1.80 \text{ g } CO_2} = 68 \text{ L } CO_2$

63. $\frac{0.702 \text{ g } CO_2}{1 \text{ min}} \times 10 \text{ min} \times \frac{1 \text{ mol } CO_2}{44.01 \text{ g } CO_2} \times \frac{4 \text{ mol } KO_2}{4 \text{ mol } CO_2} \times \frac{71.10 \text{ g } KO_2}{1 \text{ mol } KO_2}$

$= 11.3 \text{ g } KO_2$

65. a. $Cl_2(g) + 3F_2(g) \longrightarrow 2ClF_3(g)$

b. $1.75 \text{ mol } Cl_2 \times \frac{2 \text{ mol } ClF_3}{1 \text{ mol } Cl_2} = 3.50 \text{ mol } ClF_3$

$3.68 \text{ mol } F_2 \times \frac{2 \text{ mol } ClF_3}{3 \text{ mol } F_2} = 2.45 \text{ mol } ClF_3$; F_2 is limiting

c. 2.45 mol ClF_3

d. $2.45 \text{ mol } ClF_3 \times \frac{1 \text{ mol } Cl_2}{2 \text{ mol } ClF_3} = 1.22 \text{ mol } Cl_2$

excess = 1.75 mol - 1.22 mol = 0.53 mol

67. $3Al(s) + 3NH_4ClO_4(s) \longrightarrow Al_2O_3(s) + AlCl_3(s) + 3NO(g) + 6H_2O(g)$

a. $5.75 \text{ g Al} \times \frac{1 \text{ mol Al}}{26.98 \text{ g Al}} \times \frac{1 \text{ mol } AlCl_3}{3 \text{ mol Al}} \times \frac{133.33 \text{ g } AlCl_3}{1 \text{ mol } AlCl_3}$

$= 9.47 \text{ g } AlCl_3$

$7.32 \text{ g } NH_4ClO_4 \times \frac{1 \text{ mol } NH_4ClO_4}{117.49 \text{ g } NH_4ClO_4} \times \frac{1 \text{ mol } AlCl_3}{3 \text{ mol } NH_4ClO_4}$

$\times \frac{133.33 \text{ g } AlCl_3}{1 \text{ mol } AlCl_3} = 2.77 \text{ g } AlCl_3$

NH_4ClO_4 is limiting; theoretical yield = 2.77 g $AlCl_3$

b. $\frac{1.87 \text{ g}}{2.77 \text{ g}} \times 100\% = 67.5\%$

69. $100 \text{ L} \times \frac{0.695 \text{ g } NH_3}{1 \text{ L}} \times \frac{1 \text{ mol } NH_3}{17.03 \text{ g } NH_3} \times \frac{4 \text{ mol NO}}{4 \text{ mol } NH_3} \times \frac{2 \text{ mol } NO_2}{2 \text{ mol NO}}$

$\times \frac{2 \text{ mol } HNO_3}{3 \text{ mol } NO_2} \times \frac{63.02 \text{ g } HNO_3}{1 \text{ mol } HNO_3} \times (0.75)^3 = 72 \text{ g } HNO_3$

71. Theor. yield aspirin = 25.0 g/0.650 = 38.5 g

Mass SA used = $38.5 \text{ g asp} \times \frac{1 \text{ mol asp}}{180.15 \text{ g asp}} \times \frac{2 \text{ mol SA}}{2 \text{ mol asp}} \times \frac{138.12 \text{ g}}{1 \text{ mol SA}}$

= 29.5 g salicylic acid

Mass AA used = $38.5 \text{ g asp} \times \frac{1 \text{ mol asp}}{180.15 \text{ g asp}} \times \frac{1 \text{ mol AA}}{2 \text{ mol asp}}$

$\times \frac{102.08 \text{ g AA}}{1 \text{ mol AA}} \times 1.500 = 16.4$ g acetic anhydride

73. Consider 1000 O^{2-} ions; total negative charge = 2000

Let x = number of Ni^{2+} ions; 970 - x = no. Ni^{3+} ions

2x + 3(970 - x) = 2000; x = 910

$\% \ Ni^{2+} = \frac{910}{970} \times 100\% = 93.8\%$; 6.2% Ni^{3+}

75. see p. 67

77. 5.00×10^{24} molecules x $\frac{1 \text{ mol}}{6.022 \times 10^{23} \text{ molecules}}$ x $\frac{153.81 \text{ g}}{1 \text{ mol}}$

x $\frac{1 \text{ cm}^3}{1.594 \text{ g}}$ = 801 cm^3

79. The one with the smallest percent of nitrogen will cost the least

urea % N = $\frac{28.02}{60.06}$ x 100% = 46.65%

ammonia % N = $\frac{14.01}{17.03}$ x 100% = 82.27

NH_4NO_3 % N = $\frac{28.02}{80.05}$ x 100% = 35.00%

guanidine % N = $\frac{42.03}{59.08}$ x 100% = 71.14%

ammonium nitrate costs the least

81. 24.30 g/mol = 0.0272 $\mathcal{M}$; $\mathcal{M}$ = 893 g/mol

82. 107.9 g x $\frac{1 \text{ cm}^3}{10.5 \text{ g}}$ x $\frac{10^{21} \text{ nm}^3}{1 \text{ cm}^3}$ x $\frac{4 \text{ atoms}}{(0.409 \text{ nm})^3}$ = 6.01×10^{23} atoms

83. mass CaO = 4.832 g $Ca(OH)_2$ x $\frac{1 \text{ mol } Ca(OH)_2}{74.10 \text{ g } Ca(OH)_2}$ x $\frac{1 \text{ mol CaO}}{1 \text{ mol } Ca(OH)_2}$

x $\frac{56.08 \text{ g CaO}}{1 \text{ mol CaO}}$ = 3.657 g CaO

mass Ca to CaO = 3.657 g CaO x $\frac{40.08 \text{ g Ca}}{56.08 \text{ g CaO}}$ = 2.614 g Ca

mass Ca_3N_2 = 2.411 g Ca x $\frac{148.26 \text{ g } Ca_3N_2}{120.24 \text{ g Ca}}$ = 2.973 g Ca_3N_2

84. Let x = mass KBr

3.595 g - x + $\frac{x(74.55)}{119.00}$ = 3.129 g; x = 1.25 g

% KBr = $\frac{1.25 \text{ g}}{3.595 \text{ g}}$ x 100% = 34.7%

85. a. 1st oxide: 2.573 g V + 2.016 g O

0.05051 mol V, 0.1260 mol O

0.1260/0.05051 = 2.50; V_2O_5

2nd oxide: 2.573 g V + 1.209 g O

0.05051 mol V, 0.07556 mol O

0.07556/0.05051 = 1.50; V_2O_3

b. $2.016 \text{ g O} \times \frac{18.02 \text{ g } H_2O}{16.00 \text{ g O}} = 2.271 \text{ g } H_2O$

86. $C_{17}H_{21}O_4N \longrightarrow 17\ CO_2$; $C_{12}H_{22}O_{11} \longrightarrow 12\ CO_2$

Let x = mass cocaine

$$\frac{x(748.17)}{(303.35)} \times \frac{1}{1.80} + (1.00 - x)\frac{(528.12)}{(342.30)} \times \frac{1}{1.80} = 1.00$$

1.37x + 0.857 - 0.857x = 1.00

x = 0.28; 28%

CHAPTER 4
Reactions in Aqueous Solution

LECTURE NOTES

This chapter deals with three types of reactions that students will meet with again, in lecture and laboratory: precipitation, acid-base, and oxidation-reduction. Emphasis is placed on the net ionic equations written to represent these reactions and the mass-mole relationships derived from these equations. Students have a lot of trouble writing equations from scratch; balancing redox equations is relatively easy because they follow a set of rules.

Allow three lectures for this chapter. The first deals with writing net ionic equations for precipitation and acid-base reactions. The second lecture covers oxidation number and the balancing of redox equations. The final lecture covers the concept of molarity, solution stoichiometry, and volumetric analysis.

Note that:

1. Students must learn the solubility rules to write equations for precipitation reactions. By the same token, they must know the strong acids and strong bases if they are to write equations for acid-base reactions.

2. Only molecular weak acids and weak bases are considered here; acidic and basic ions are covered in Chapter 13.

3. The half-equation method is the only one described for balancing redox equations. It has the advantage that it gets students into the habit of breaking down the reaction into an oxidation and a reduction. This will come in handy in Chapter 17.

LECTURE 1

I <u>Precipitation Reactions</u>

A. <u>Solubility rules</u> (Table 4.1)

Use in predicting results of precipitation reactions:

1. Mix solutions of $Ba(NO_3)_2$ and Na_2CO_3. What happens?

Ions present: Ba^{2+}, NO_3^-, Na^+, CO_3^{2-}

Possible precipitates: $BaCO_3$, $NaNO_3$

According to solubility rules, $BaCO_3$ is insoluble:

$Ba^{2+}(aq) + CO_3^{2-}(aq) \longrightarrow BaCO_3(s)$

2. Mix solutions of $BaCl_2$, NaOH

   Ions present: Ba^{2+}, Cl^-, Na^+, OH^-

   Possible precipitates: $Ba(OH)_2$, NaCl

   Both are soluble; no reaction

II <u>Acids and Bases</u>

A. <u>Acid</u> forms H^+ ion in water

Strong acids are completely dissociated in water (HCl, HBr, HI, HNO_3, $HClO_4$, H_2SO_4)

$HCl(aq) \longrightarrow H^+(aq) + Cl^-(aq)$

Reaction goes to completion; no HCl molecules in solution

Weak acids are partially dissociated in water

$HF(aq) \rightleftharpoons H^+(aq) + F^-(aq)$

Hydrofluoric acid contains a mixture of HF molecules, H^+ and F^- ions.

B. <u>Base</u> forms OH^- in solution

Strong base is completely dissociated in water (Group 1 and heavier Group 2 hydroxides)

$Ca(OH)_2(s) \longrightarrow Ca^{2+}(aq) + 2\ OH^-(aq)$

Weak base is partially dissociated to form OH^- ions

$NH_3(aq) + H_2O \rightleftharpoons NH_4^+(aq) + OH^-(aq)$

C. <u>Acid-Base reactions</u>

1. Strong acid + strong base: HCl + $Ca(OH)_2$

   $H^+(aq) + OH^-(aq) \longrightarrow H_2O$

2. Strong base + weak acid: HF + $Ca(OH)_2$

   $HF(aq) + OH^-(aq) \longrightarrow F^-(aq) + H_2O$

3. Strong acid + weak base: HCl, NH_3

$$H^+(aq) + NH_3(aq) \longrightarrow NH_4^+(aq)$$

LECTURE 2

I Redox Reactions

A. Oxidation number; "pseudocharge" assigned according to arbitrary rules.

1. Oxidation number of element in elementary substance (e.g., F_2, O_2) is zero.

2. Oxidation number of element in monatomic ion is the charge of that ion. Oxidation number of iron is +2 in Fe^{2+}, +3 in Fe^{3+}.

3. Oxidation number of Group 1 elements in their compounds is +1; +2 for Group 2; -1 for fluorine.

   Oxid. no. H is almost always +1

   Oxid. no. of O is ordinarily -2

4. Sum of oxidation numbers of all atoms in molecule = 0. In polyatomic ion, the sum is the charge of the ion.

   H_2SO_4: +2 + (oxid. no. S) + 4(-2) = 0; oxid no S = +6

   $Cr_2O_7^{2-}$: 2(oxid. no. Cr) + 7(-2) = -2; oxid no Cr = +6

B. Oxidation = increase in oxid. no.; reduction = decrease in oxidation number.

$$HCl(g) + HNO_3(l) \longrightarrow NO_2(g) + \tfrac{1}{2}Cl_2(g) + H_2O(l)$$

HCl is oxidized; oxid. no. Cl increases from -1 to 0

HNO_3 is reduced; oxid. no. N decreases from +5 to +4

II Balancing Redox Equations

A. $Cr_2O_7^{2-}(aq) + I^-(aq) \longrightarrow Cr^{3+}(aq) + I_2(s)$ (acid)

1. Split into two half-equations

   oxidation: $I^-(aq) \longrightarrow I_2(s)$

   reduction: $Cr_2O_7^{2-}(aq) \longrightarrow Cr^{3+}(aq)$

2. Balance half-equations separately. First, balance element whose oxidation number changes. Then balance oxygen, then hydrogen. Finally, balance charge by adding electrons.

a. $2I^-(aq) \longrightarrow I_2(s)$

$2I^-(aq) \longrightarrow I_2(s) + 2e^-$

b. $Cr_2O_7^{2-}(aq) \longrightarrow 2Cr^{3+}(aq)$

$Cr_2O_7^{2-}(aq) \longrightarrow 2Cr^{3+}(aq) + 7H_2O$

$Cr_2O_7^{2-}(aq) + 14H^+(aq) \longrightarrow 2Cr^{3+}(aq) + 7H_2O$

$Cr_2O_7^{2-}(aq) + 14H^+(aq) + 6e^-(aq) \longrightarrow 2Cr^{3+}(aq) + 7H_2O$

3. Combine half-equations so that electrons cancel.

$$3[\ 2I^-(aq) \longrightarrow I_2(s) + 2e^-]$$

$$\underline{Cr_2O_7^{2-}(aq) + 14H^+(aq) + 6e^- \longrightarrow 2Cr^{3+}(aq) + 7H_2O}$$

$$6I^-(aq) + Cr_2O_7^{2-}(aq) + 14H^+(aq) \longrightarrow 3I_2(s) + 2Cr^{3+}(aq) + 7H_2O$$

B. To balance in basic solution, "neutralize" H^+ ions by adding OH^- ions.

With above equation, add 14 OH^- ions to both sides, obtaining, after simplification:

$$6I^-(aq) + Cr_2O_7^{2-}(aq) + 7H_2O \longrightarrow 3I_2(s) + 2Cr^{3+}(aq) + 14\ OH^-$$

LECTURE 3

I Molarity

The amount of solute in a solution sample depends upon two factors: the volume of solution and the concentration of solute.

$$\text{Molarity (M)} = \frac{\text{no. moles solute}}{\text{no. liters solution}}$$

1. What volume of 12 M HCl must be taken to obtain 0.10 mol of HCl?

$$\text{Volume} = 0.10 \text{ mol HCl} \times \frac{1 \text{ L}}{12 \text{ mol HCl}} = 0.0083 \text{ L} = 8.3 \text{ mL}$$

2. What mass of NaOH is contained in 125 mL of 6.00 M NaOH?

$$125 \text{ mL} \times \frac{1 \text{ L}}{1000 \text{ mL}} \times \frac{6.00 \text{ mol}}{1 \text{ L}} \times \frac{40.00 \text{ g}}{1 \text{ mol}} = 30.0 \text{ g NaOH}$$

3. What are the molarities of Al^{3+} and SO_4^{2-} in 0.100 M $Al_2(SO_4)_3$?

$$Al_2(SO_4)_3(s) \longrightarrow 2Al^{3+}(aq) + 3SO_4^{2-}(aq)$$

$[Al^{3+}] = 0.200$ M $[SO_4^{2-}] = 0.300$ M

II Solution Stoichiometry

A. $Cu^{2+}(aq) + 2\ OH^-(aq) \longrightarrow Cu(OH)_2(s)$

What volume of 0.200 M $CuSO_4$ solution is required to react with 50.0 mL of 0.100 M NaOH?

$$n\ OH^- = 0.0500\ L \times \frac{0.100\ mol\ NaOH}{1\ L} \times \frac{1\ mol\ OH^-}{1\ mol\ NaOH}$$

$$= 0.00500\ mol\ OH^-$$

$$n\ Cu^{2+} = 0.00500\ mol\ OH^- \times \frac{1\ mol\ Cu^{2+}}{2\ mol\ OH^-} = 0.00250\ mol\ Cu^{2+}$$

$$V = 0.00250\ mol\ Cu^{2+} \times \frac{1\ mol\ CuSO_4}{1\ mol\ Cu^{2+}} \times \frac{1\ L}{0.200\ mol\ CuSO_4}$$

$$= 0.0125\ L = 12.5\ mL$$

B. $Cr_2O_7^{2-}(aq) + 6I^-(aq) + 14H^+(aq) \longrightarrow 3I_2(s) + 2Cr^{3+}(aq) + 7H_2O$

Suppose 22.0 mL of 0.150 M $K_2Cr_2O_7$ is required to react with a sample weighing 5.00 g. Percent of I^- in sample?

$$n\ Cr_2O_7^{2-} = 0.0220\ L \times \frac{0.150\ mol\ K_2Cr_2O_7}{1\ L} \times \frac{1\ mol\ Cr_2O_7^{2-}}{1\ mol\ K_2Cr_2O_7}$$

$$= 0.00330\ mol\ Cr_2O_7^{2-}$$

$$mass\ I^- = 0.00330\ mol\ Cr_2O_7^{2-} \times \frac{6\ mol\ I^-}{1\ mol\ Cr_2O_7^{2-}} \times \frac{126.9\ g\ I^-}{1\ mol\ I^-}$$

$$= 2.51\ g\ I^-$$

$$\%\ I^- = \frac{2.51\ g}{5.00\ g} \times 100 = 50.2$$

DEMONSTRATIONS

1. Precipitation of lead iodide: Test. Dem. 45
2. Precipitation of mercury(II) thiocyanate (Pharaoh's Serpents) Test. Dem. 29, 94

3. Classical properties of acids and bases: Shak. 3 58
4. Strength of acids: Shak. 3 136
5. Reaction of barium hydroxide with sulfuric acid: Test. Dem. 163
6. End point of acid-base titration: Shak. 3 152
7. Neutralizing antacids: Shak. 3 162
8. Oxidation of zinc: Test. Dem. 154
9. Formation of Sn, Pb, Ag trees: Test. Dem. 53, 127
*10. Reaction of zinc with sulfur: Shak. 1 53
11. Reaction of sodium peroxide with aluminum: Shak. 1 59
12. Breathalyzer: J. Chem. Educ. 67 263 (1990)

* See also Shakhashiri Videotapes, Demonstration 23

QUIZZES

Quiz 1

1. Write a net ionic equation for the reaction between:
   a. solutions of HCl and $Ba(OH)_2$
   b. solutions of Na_2SO_4 and $Ba(OH)_2$
   List the ions present in each solution before reaction
2. Using the equation in 1(a), calculate the volume of 0.100 M $Ba(OH)_2$ required to react with 22.0 mL of 0.100 M HCl.

Quiz 2

1. Balance the following redox equation:

   $MnO_4^-(aq) + Fe^{2+}(aq) \longrightarrow Mn^{2+}(aq) + Fe^{3+}(aq)$ (acid)

2. Using the balanced equation in (1), calculate the concentration of a solution of $KMnO_4$ if 25.0 mL of it is required to react with 20.0 mL of 0.504 M $FeSO_4$ solution.

Quiz 3

1. List six strong acids; six strong bases.
2. Write a balanced net ionic equation for
   a. the reaction between solutions of $Pb(NO_3)_2$ and $FeCl_3$.
   b. the reaction between ClO_4^- and SO_3^{2-} in acidic solution; products include Cl^- and SO_4^{2-} ions.
3. What are the molarities of ions in

a. 0.25 M K_2SO_4?   b. 0.13 M $Al(NO_3)_3$?

Quiz 4

When calcium oxalate is treated with an acidic solution of potassium permanganate, the following reaction occurs:

$$CaC_2O_4(s) + MnO_4^-(aq) \longrightarrow Ca^{2+}(aq) + CO_2(g) + Mn^{2+}(aq)$$

1. Balance the equation.
2. Calculate the percent of calcium oxalate in a sample if 25.0 ml of 0.1000 M $KMnO_4$ is required to react with 1.000 g of sample.

Quiz 5

1. Give the formulas of three insoluble chlorides; three soluble carbonates.

2. Write an equation for the reaction between solutions of nitrous acid, HNO_2, and calcium hydroxide. What volume of 0.0512 M $Ca(OH)_2$ is required to react with 0.200 g of HNO_2?

Answers

Quiz 1

1. ions present: H^+, Cl^-, Ba^{2+}, OH^-

   $H^+(aq) + OH^-(aq) \longrightarrow H_2O$

2. ions present: Na^+, SO_4^{2-}, Ba^{2+}, OH^-

   $Ba^{2+}(aq) + SO_4^{2-}(aq) \longrightarrow BaSO_4(s)$

2. 11.0 mL

Quiz 2

1. $MnO_4^-(aq) + 5Fe^{2+}(aq) + 8H^+(aq) \longrightarrow Mn^{2+}(aq) + 5Fe^{3+}(aq) + 4H_2O$
2. 0.0806 M

Quiz 3

1. HCl, HBr, HI, HNO_3, $HClO_4$, H_2SO_4; LiOH, NaOH, KOH, $Ca(OH)_2$ $Sr(OH)_2$, $Ba(OH)_2$
2. a. $Pb^{2+}(aq) + 2Cl^-(aq) \longrightarrow PbCl_2(s)$

   b. $ClO_4^-(aq) + 4SO_3^{2-}(aq) \longrightarrow Cl^-(aq) + 4SO_4^{2-}(aq)$
3. a. 0.25 M SO_4^{2-}, 0.50 M K^+ b. 0.13 M Al^{3+}, 0.39 M NO_3^-

Quiz 4

1. $5CaC_2O_4(s) + 2MnO_4^-(aq) + 16H^+(aq) \rightarrow 5Ca^{2+}(aq) + 2Mn^{2+}(aq) + 10\ CO_2(g) + 8H_2O$

2. 80.1%

Quiz 5

1. $PbCl_2$, Hg_2Cl_2, $AgCl$; Na_2CO_3, K_2CO_3, $(NH_4)_2CO_3$

2. $HNO_2(aq) + OH^-(aq) \rightarrow NO_2^-(aq) + H_2O$; 41.5 mL

PROBLEMS

1. a. $CaCl_2$; soluble b. $(NH_4)_2S$; soluble
   c. $BaSO_4$; insoluble d. $Fe(OH)_2$; insoluble

3. a. Na_2CO_3 b. Na_2S or H_2S c. NaOH

5. a. $Ni^{2+}(aq) + 2\ OH^-(aq) \rightarrow Ni(OH)_2(s)$
   b. $Mg^{2+}(aq) + 2\ OH^-(aq) \rightarrow Mg(OH)_2(s)$

7. a. no precipitate
   b. $Pb^{2+}(aq) + 2Cl^-(aq) \rightarrow PbCl_2(s)$
   c. $Ba^{2+}(aq) + SO_4^{2-}(aq) \rightarrow BaSO_4(s)$
   d. no precipitate
   e. $2As^{3+}(aq) + 3S^{2-}(aq) \rightarrow As_2S_3(s)$

9. a. $Ba^{2+}(aq) + CO_3^{2-}(aq) \rightarrow BaCO_3(s)$
   b. $Zn^{2+}(aq) + S^{2-}(aq) \rightarrow ZnS(s)$
   c. $Ag^+(aq) + Cl^-(aq) \rightarrow AgCl(s)$
   d. no precipitate

11. a. H^+ b. HF c. H^+ d. H_2SO_3 e. $HC_3H_5O_2$

13. a. OH^- b. $C_2H_5NH_2$ c. NH_3 d. OH^-

15. a. strong acid b. strong base c. weak acid d. weak base

17. a. $HC_4H_7O_2(aq) + OH^-(aq) \rightarrow H_2O + C_4H_7O_2^-(aq)$

b. $NH_3(aq) + H^+(aq) \longrightarrow NH_4^+(aq)$

c. $HF(aq) + OH^-(aq) \longrightarrow H_2O + F^-(aq)$

19. a. correct

b. $H^+(aq) + CH_3NH_2(aq) \longrightarrow CH_3NH_3^+(aq)$

c. $H^+(aq) + NH_3(aq) \longrightarrow NH_4^+(aq)$

d. correct e. correct

f. $HF(aq) + OH^-(aq) \longrightarrow H_2O + F^-(aq)$

21. a. HCO_3^- H(+1), O(-2), C(+4)

b. $MgSO_4$ Mg(+2), O(-2), S(+6)

c. SF_6 S(+6), F(-1)

d. HIO_3 H(+1), O(-2), I(+5)

e. Na_2MoO_4 Na(+1), O(-2), Mo(+6)

23. a. O(-2), Sb(+5)

b. H(+1), O(-2), P(+3)

c. F(-1), Ru(+5)

d. H(+1), O(-2), C(-1)

25. a. oxidation (oxid. no. Ca: $0 \rightarrow +2$)

b. oxidation (oxid. no. O: $-2 \rightarrow 0$)

c. reduction (oxid. no. N: $+5 \rightarrow +2$)

d. reduction (oxid. no. Al: $+3 \rightarrow +1$)

27. a. Mg is oxidized, O_2 reduced; Mg reducing agent, O_2 oxidizing agent

b. $Cr_2O_7^{2-}$ reduced, Sn^{2+} oxidized; $Cr_2O_7^{2-}$ oxidizing agent, Sn^{2+} reducing agent

29. a. $2Mg(s) + O_2(g) \longrightarrow 2MgO(s)$

b. $Cr_2O_7^{2-}(aq) + 14H^+(aq) + 6e^- \longrightarrow 2Cr^{3+}(aq) + 7H_2O$

$\underline{3[Sn^{2+}(aq) \longrightarrow Sn^{4+}(aq) + 2e^-]}$

$Cr_2O_7^{2-}(aq) + 14H^+(aq) + 3Sn^{2+}(aq) \rightarrow 2Cr^{3+}(aq) + 3Sn^{4+}(aq) + 7H_2O$

31. a. $Hg^{2+}(aq) + Cu(s) \rightarrow Hg(l) + Cu^{2+}(aq)$

    b. $3Zn(s) + 2VO_3^-(aq) + 12H^+(aq) \rightarrow 3Zn^{2+}(aq) + 2V^{2+}(aq) + 6H_2O$

    c. $3H_2O_2(aq) + Cr_2O_7^{2-}(aq) + 8H^+(aq) \rightarrow 3\ O_2(g) + 2Cr^{3+}(aq) + 7H_2O$

    d. $MnO_2(s) + 4H^+(aq) + 2Cl^-(aq) \rightarrow Mn^{2+}(aq) + Cl_2(g) + 2H_2O$

    e. $IO_3^-(aq) + 6H^+(aq) + 8I^-(aq) \rightarrow 3I_3^-(aq) + 3H_2O$

33. a. $P_4(s) + 6H_2O + 10HClO(aq) \rightarrow 4H_3PO_4(aq) + 10\ H^+(aq) + 10\ Cl^-(aq)$

    b. $3Te(s) + 4NO_3^-(aq) + 4H^+(aq) \rightarrow 3TeO_2(s) + 4NO(g) + 2H_2O$

    c. $3Br_2(aq) + I^-(aq) + 3H_2O \rightarrow 6Br^-(aq) + IO_3^-(aq) + 6H^+(aq)$

35. a. $3ClO^-(aq) + 2CrO_2^-(aq) + 2\ OH^-(aq) \rightarrow 3Cl^-(aq) + 2CrO_4^{2-}(aq) + H_2O$

    b. $2Al(s) + 6H_2O + 2\ OH^-(aq) \rightarrow 2Al(OH)_4^-(aq) + 3H_2(g)$

    c. $2Ni^{2+}(aq) + 6\ OH^-(aq) + Br_2(l) \rightarrow 2NiO(OH)(s) + 2Br^-(aq) + 2H_2O$

37. n = 0.425 L x 0.628 mol/L = 0.267 mol

    a. $\mathcal{M}$ = 194.20 g/mol. Dissolve 0.267 mol (51.9 g) of K_2CrO_4 in enough water to form 425 mL of solution.

    b. $\mathcal{M}$ = 149.9 g/mol. Dissolve 0.267 mol (40.0 g) of NaI in enough water to form 425 mL of solution.

    c. $\mathcal{M}$ = 180.16 g/mol. Dissolve 0.267 mol (48.1 g) of $C_6H_{12}O_6$ in enough water to form 425 mL of solution.

39. a. 12.45×10^{-3} L x 3.00 mol/L = 0.0374 mol

    b. $0.800 \text{ mol} \times \frac{1 \text{ L}}{3.00 \text{ mol}} = 0.267 \text{ L}$

41. a. $\mathcal{M}$ = 105.99 g/mol; 3.58 g, 0.250 L, <u>0.135 mol/L</u>

    b. $\mathcal{M}$ = 32.04 g/mol ; <u>96.1 g</u>, 0.500 L, 6.00 mol/L

c. $\mathcal{M}$ = 261. 3 g/mol; 5.89 g, 0.0186 L, 1.21 mol/L

d. $\mathcal{M}$ = 342.17 g/mol; 16.3 g, 0.455 L, 0.105 mol/L

43. a. $CaBr_2$; 0.25 mol x 3 = 0.75 mol

b. $MgSO_4$; 0.25 mol x 2 = 0.50 mol

c. $Fe(NO_3)_3$; 0.25 mol x 4 = 1.0 mol

d. $NiSO_4$; 0.25 mol x 2 = 0.50 mol

45. a. $n\ Ag^+ = 0.0220\ L \times 0.130\ mol/L = 0.00286\ mol = n\ Cl^-$

$$n\ ScCl_3 = 0.00286\ mol\ Cl^- \times \frac{1\ mol\ ScCl_3}{3\ mol\ Cl^-} = 9.53 \times 10^{-4}\ mol$$

$$V = 9.53 \times 10^{-4}\ mol \times \frac{1\ L}{3.85 \times 10^{-2}\ mol} = 2.48 \times 10^{-2}\ L$$

b. 0.00286 mol x 143.4 g/mol = 0.410 g

47. a. $Pb^{2+}(aq) + SO_4^{2-}(aq) \longrightarrow PbSO_4(s)$

$n\ SO_4^{2-} = 0.0250\ L \times 0.0832\ mol/L = 2.08 \times 10^{-3}\ mol$

$= n\ Pb(NO_3)_2$

$$V = 2.08 \times 10^{-3}\ mol \times \frac{1\ L}{0.100\ mol} = 2.08 \times 10^{-2}\ L$$

b. $Pb^{2+}(aq) + 2Cl^-(aq) \longrightarrow PbCl_2(s)$

$n\ Cl^- = 0.0558\ L \times 0.222\ mol/L = 0.0124\ mol$

$$n\ Pb(NO_3)_2 = 0.0124\ mol\ Cl^- \times \frac{1\ mol\ Pb(NO_3)_2}{2\ mol\ Cl^-} = 0.00620\ mol$$

$$V = 0.00620\ mol \times \frac{1\ L}{0.100\ mol} = 6.20 \times 10^{-2}\ L$$

c. $Pb^{2+}(aq) + CrO_4^{2-}(aq) \longrightarrow PbCrO_4(s)$

$n\ CrO_4^{2-} = 0.0187\ L \times 0.389\ mol/L = 0.00727\ mol$

$= n\ Pb(NO_3)_2$

$$V = 0.00727\ mol \times \frac{1\ L}{0.100\ mol} = 7.27 \times 10^{-2}\ L$$

49. $H^+(aq) + OH^-(aq) \longrightarrow H_2O$

$n\ H^+ = 0.0384\ L \times 0.215\ mol/L = 0.00826\ mol = n\ OH^-$

$n\ Ba(OH)_2 = 0.00826\ mol\ OH^- \times \frac{1\ mol\ Ba(OH)_2}{2\ mol\ OH^-} = 0.00413\ mol$

$M\ Ba(OH)_2 = 0.00413\ mol/0.0200\ L = 0.206\ mol/L$

51. a. $H^+(aq) + OH^-(aq) \longrightarrow H_2O$

$n\ OH^- = 0.0300\ L \times 0.278\ mol/L = 0.00834\ mol = n\ HCl$

$V = 0.00834\ mol \times \frac{1\ L}{0.250\ mol} = 0.0334\ L$

b. $n\ Sr(OH)_2 = 0.0176\ L \times 0.0162\ mol/L = 2.85 \times 10^{-4}\ mol$

$n\ OH^- = 2 \times 2.85 \times 10^{-4}\ mol = 5.70 \times 10^{-4}\ mol = n\ HCl$

$V = 5.70 \times 10^{-4}\ mol \times \frac{1\ L}{0.250\ mol} = 2.28 \times 10^{-3}\ L$

c. $NH_3(aq) + H^+(aq) \longrightarrow NH_4^+(aq)$

$n\ NH_3 = 15.0\ mL \times \frac{0.958\ g}{1\ mL} \times \frac{1\ mol}{17.03\ g} \times 0.100 = 0.0844\ mol$

$V = 0.0844\ mol \times \frac{1\ L}{0.250\ mol} = 0.338\ L$

53. a. $I_2(s) + 2S_2O_3^{2-}(aq) \longrightarrow 2I^-(aq) + S_4O_6^{2-}(aq)$

b. $n\ I_2 = 10.0\ g \times \frac{1\ mol}{253.8\ g} = 0.0394\ mol$

$n\ Na_2S_2O_3 = 2 \times 0.0394\ mol = 0.0788\ mol$

$V = 0.0788\ mol \times \frac{1\ L}{0.125\ mol} = 0.630\ L$

55. a. $Cu(s) + 2NO_3^-(aq) + 4H^+(aq) \longrightarrow Cu^{2+}(aq) + 2NO_2(g) + 2H_2O$

b. $n\ Cu = 5.87\ g \times \frac{1\ mol}{63.55\ g} = 9.24 \times 10^{-2}\ mol$

$n\ H^+ = 4 \times 9.24 \times 10^{-2}\ mol = 3.70 \times 10^{-1}\ mol$

$V = 3.70 \times 10^{-1}\ mol \times \frac{1\ L}{16.0\ mol} = 2.31 \times 10^{-2}\ L$

57. $Cr_2O_7^{2-}(aq) + 6Fe^{2+}(aq) + 14H^+(aq) \longrightarrow 2Cr^{3+}(aq) + 6Fe^{3+}(aq) + 7H_2O$

a. $n\ Fe^{2+} = 0.0293\ L \times 0.0325\ mol/L = 9.52 \times 10^{-4}\ mol$

$n\ K_2Cr_2O_7 = 9.52 \times 10^{-4}\ mol\ Fe^{2+} \times \frac{1\ mol\ K_2Cr_2O_7}{6\ mol\ Fe^{2+}}$

$= 1.59 \times 10^{-4}\ mol$

$M\ K_2Cr_2O_7 = \frac{1.59 \times 10^{-4}\ mol}{2.38 \times 10^{-2}\ L} = 6.68 \times 10^{-3}\ mol/L$

b. $n\ H^+ = 9.52 \times 10^{-4}\ mol\ Fe^{2+} \times \frac{14\ mol\ H^+}{6\ mol\ Fe^{2+}} = 2.22 \times 10^{-3}\ mol\ H^+$

$V = 2.22 \times 10^{-3}\ mol \times \frac{1\ L}{0.100\ mol} = 2.22 \times 10^{-2}\ L$

59. $H^+(aq) + OH^-(aq) \longrightarrow H_2O$

$n\ H^+ = 0.02441\ L \times 0.1645\ mol/L = 4.015 \times 10^{-3}\ mol = n\ NaOH$

$M\ NaOH = \frac{4.015 \times 10^{-3}\ mol}{2.000 \times 10^{-2}\ L} = 0.2008\ mol/L$

61. $n\ OH^- = 0.0103\ L \times 0.250\ mol/L = 2.58 \times 10^{-3}\ mol = n\ C_6H_8O_6$

$\%\ C_6H_8O_6 = 2.58 \times 10^{-3}\ mol \times \frac{176.12\ g}{1\ mol} \times \frac{100}{0.518\ g} = 87.7$

63. $n\ TA = 12.0\ g \times \frac{1\ mol}{150.09\ g} = 0.0800\ mol$

$n\ KOH = 0.0800\ mol\ TA \times \frac{2\ mol\ KOH}{1\ mol\ TA} = 0.160\ mol\ KOH$

$mass\ KOH = 0.160\ mol \times \frac{56.11\ g}{1\ mol} = V \times \frac{1.045\ g}{1\ cm^3} \times 0.0500$

$V = 172\ cm^3$

65. $n\ Fe^{2+} = 0.100\ g \times 0.9978 \times \frac{1\ mol}{55.85\ g} = 1.79 \times 10^{-3}\ mol$

$n\ Sn^{2+} = 1.79 \times 10^{-3}\ mol/2 = 8.95 \times 10^{-4}\ mol$

$M = \frac{8.95 \times 10^{-4}\ mol}{9.47 \times 10^{-3}\ L} = 0.0945\ mol/L$

67. $2Cr_2O_7^{2-}(aq) + C_2H_5OH(aq) + 16H^+(aq) \longrightarrow 4Cr^{3+}(aq) + 2CO_2(g) + 11\ H_2O$

$n\ Cr_2O_7^{2-} = 0.04502\ L \times 0.05000\ mol/L = 2.251 \times 10^{-3}\ mol$

$n\ C_2H_5OH = 2.251 \times 10^{-3}\ mol/2 = 1.126 \times 10^{-3}\ mol$

$\%\ C_2H_5OH = 1.126 \times 10^{-3}\ mol \times \frac{46.07\ g}{1\ mol} \times \frac{100\%}{50.00\ g} = 0.1037\%$; yes

69. RNH_2, R_2NH, R_3N

71. See p. 96

73. a. precipitation b. acid-base c. redox d. redox

75. $5Fe^{2+}(aq) + MnO_4^-(aq) + 8H^+(aq) \longrightarrow 5Fe^{3+}(aq) + Mn^{2+}(aq) + 4H_2O$

$n\ MnO_4^- = 3.23 \times 10^{-2}\ L \times 0.002100\ mol/L = 6.78 \times 10^{-5}\ mol$

$n\ Fe^{2+} = 5 \times 6.78 \times 10^{-5}\ mol = 3.39 \times 10^{-4}\ mol$

$\%\ Fe = 3.39 \times 10^{-4}\ mol \times \frac{55.85\ g}{1\ mol} \times \frac{100\%}{5.00\ g} = 0.379\%$

77. $5CaC_2O_4(s) + 2MnO_4^-(aq) + 16H^+(aq) \longrightarrow 10\ CO_2(g) + 2Mn^{2+}(aq) + 5Ca^{2+}(aq) + 8H_2O$

$n\ MnO_4^- = 0.0262\ L \times 0.0946\ mol/L = 2.48 \times 10^{-3}\ mol$

$n\ CaC_2O_4 = 2.48 \times 10^{-3}\ mol \times 5/2 = 6.20 \times 10^{-3}\ mol$

mass $Ca = 6.20 \times 10^{-3}\ mol \times 40.08\ g/mol = 0.248\ g$; yes

78. $n\ Mg(OH)_2 = 0.330\ g \times 0.410 \times \frac{1\ mol}{58.32\ g} = 0.00232\ mol$

$n\ NaHCO_3 = 0.330\ g \times 0.362 \times \frac{1\ mol}{84.01\ g} = 0.00142\ mol$

$n\ H^+ = 2(0.00232\ mol) + 0.00142\ mol = 0.00606\ mol$

$V = 0.00606\ mol \times \frac{1\ L}{0.020\ mol} = 0.30\ L$

79. $n\ HNO_3 = 2.0 \times 10^4\ gal \times \frac{4\ qt}{1\ gal} \times \frac{1\ L}{1.057\ qt} \times \frac{1420\ g}{1\ L} \times 0.72$

$\times \frac{1\ mol}{63.02\ g} = 1.2 \times 10^6\ mol$

mass $Na_2CO_3 = 1.2 \times 10^6\ mol\ H^+ \times \frac{1\ mol\ Na_2CO_3}{2\ mol\ H^+} \times \frac{105.99\ g\ Na_2CO_3}{1\ mol\ Na_2CO_3}$

$= 6.5 \times 10^7\ g$

80. $5Fe^{2+}(aq) + MnO_4^-(aq) + 8H^+(aq) \longrightarrow 5Fe^{3+}(aq) + Mn^{2+}(aq) + 4H_2O$

1) $n\ MnO_4^- = 0.0350\ L \times 0.0280\ mol/L = 9.80 \times 10^{-4}\ mol$

$n\ Fe^{2+} = 5 \times 9.80 \times 10^{-4}\ mol = 4.90 \times 10^{-3}\ mol$

$$M\ Fe^{2+} = \frac{4.90 \times 10^{-3}\ mol}{5.000 \times 10^{-2}\ L} = 0.0980\ mol/L$$

2) $n\ MnO_4^- = 0.0480\ L \times 0.0280\ mol/L = 1.34 \times 10^{-3}\ mol$

$n\ Fe^{2+} = 5 \times 1.34 \times 10^{-3}\ mol = 6.72 \times 10^{-3}\ mol$

$n\ Fe^{3+} = 6.72 \times 10^{-3}\ mol - 4.90 \times 10^{-3}\ mol = 1.82 \times 10^{-3} mol$

$$M\ Fe^{3+} = \frac{1.82 \times 10^{-3}\ mol}{5.00 \times 10^{-2}\ L} = 0.0364\ mol/L$$

CHAPTER 5
Gases

LECTURE NOTES

The ideal gas law is at the heart of this chapter. It is unnecessary to spend much time on the relationships between variables such as Boyle's law or Charles' law. For one thing, students almost certainly got a heavy dose of problems of that type in high school. Moreover, such relationships have very limited application to chemistry.

Dalton's law is another one that students have very little trouble with. The relation between mole fraction and partial pressure, though, is important; it comes up later in the guise of Raoult's law or Henry's law. We ordinarily spend relatively little time on deviations from the ideal gas law.

The basic equation of kinetic theory is: $\frac{1}{2}mu^2$ = constant x T. We relate u to T and to molar mass (Graham's law) through this equation. We do not attempt to derive or use the expression for average velocity. The concept of a distribution of molecular velocities is introduced here and serves as background for chemical kinetics.

This chapter is readily covered in two lectures; students have very little trouble with it.

LECTURE 1

I The Ideal Gas Law

A. Variables: V = volume (liters, cm^3, m^3)
n = amount in moles
T = temperature (K)
P = pressure (atmospheres, mm Hg)

Relation between variables:

$$PV = nRT$$

where R is a constant equal to 0.0821 L·atm/mol·K

B. Initial and final state problems Use ideal gas law to find necessary relations.

A cylinder contains a gas at a pressure of 255 lb/in^2. If the valve is opened and 75% of the gas escapes, what is the final pressure?

V and T are constant, so $P_2/P_1 = n_2/n_1$; $n_2 = 0.25n_1$

$P_2 = 0.25P_1 = 0.25 \times 255\ lb/in^2 = 64\ lb/in^2$

C. Calculation of P, V, n or T

What is the pressure exerted by 15.0 mol of O_2 in a 50.0 L tank at 25°C?

$$P = \frac{nRT}{V} = \frac{(15.0\ mol)(0.0821\ L{\cdot}atm/mol{\cdot}K)(298\ K)}{50.0\ L} = 7.34\ atm$$

Note that since R = 0.0821 L•atm/mol•K, V must be in liters, P in atmospheres, T in K, n in moles.

D. Calculation of density or molar mass

$$PV = \frac{mRT}{\mathcal{M}}\ ;\quad \frac{m}{V} = d = \frac{P\mathcal{M}}{RT}$$

Density of $O_2(g)$ at 27°C, 735 mm Hg?

$$d = \frac{(735/760\ atm)(32.00\ g/mol)}{(0.0821\ L{\cdot}atm/mol{\cdot}K)(300\ K)} = 1.26\ g/L$$

A flask weighs 52.693 g empty and 53.117 g when filled with acetone vapor at 100°C and 752 mm Hg. Taking the volume of the flask to be 226.2 mL, calculate the molar mass of acetone.

$$\mathcal{M} = \frac{mRT}{PV} = \frac{(0.424\ g)(0.0821\ L{\cdot}atm/mol{\cdot}K)(373\ K)}{(752/760\ atm)(0.2262\ L)} = 58.0\ g/mol$$

LECTURE 2

I The Ideal Gas Law

A. Volumes of gases involved in reactions

$$Zn(s) + 2H^+(aq) \longrightarrow Zn^{2+}(aq) + H_2(g)$$

Mass of Zn required to form 16.0 L of $H_2(g)$ at 20°C, 735 mm Hg?

Path to follow: $V\ H_2 \rightarrow n\ H_2 \rightarrow n\ Zn \rightarrow$ mass Zn

$$n\ H_2 = \frac{PV}{RT} = \frac{(735/760\ atm)(16.0\ L)}{(0.0821\ L{\cdot}atm/mol{\cdot}K)(293\ K)} = 0.643\ mol\ H_2$$

$$\text{mass Zn} = 0.643\ mol\ H_2 \times \frac{1\ mol\ Zn}{1\ mol\ H_2} \times \frac{65.38\ g\ Zn}{1\ mol\ Zn} = 42.0\ g\ Zn$$

II Dalton's Law (Gas Mixtures)

A. $P_{tot} = P_1 + P_2 + - -$ where P_1 is the partial pressure of gas 1, etc. Most often used in collection of gases over water:

$$P_{gas} = P_{tot} - P_{H_2O}$$

where P_{H_2O} is the vapor pressure of water at the specified temperature.

B. $P_1 = X_1 P_{tot}$

where X = mole fraction of gas in mixture. Partial pressure of oxygen in air (X = 0.2095) when barometric pressure = 734 mm Hg?

$$P_{O_2} = 0.2095(734 \text{ mm Hg}) = 154 \text{ mm Hg}$$

III <u>Kinetic Theory of Gases</u>

$$E_{trans} = \tfrac{1}{2}mu^2 = C \times T$$

where m = mass of molecule, u = average speed, T = temperature in K, and C is a constant which has the same value for all gases.

A. <u>Graham's law</u> relates m and u for two different gases, sameT

$$m_2u_2^2 = m_1u_1^2$$

$$(\text{rate } 2)/(\text{rate } 1) = (\mathcal{M}_1/\mathcal{M}_2)^{\frac{1}{2}} = (\text{time } 1)/(\text{time } 2)$$

Certain gas takes 2.42 times as long to effuse as O_2 at same T, P. Molar mass of gas?

$$(\mathcal{M}/32.0 \text{ g/mol})^{\frac{1}{2}} = 2.42; \mathcal{M} = (2.42)^2 \text{x } 32.0 \text{ g/mol} = 187 \text{ g/mol}$$

IV <u>Real Gases</u>

Deviate at least slightly from the ideal gas law because of two factors:

1. Gas molecules attract one another
2. Gas molecules occupy a finite volume

Both of these factors are neglected in the ideal gas law. Both increase in importance when the molecules are close together (high P, low T)

Van der Waals equation: $(P + a/V_m^2)(V_m - b) = RT$

V_m = molar volume

a/V_m^2 corrects for attraction between molecules

b is approximately equal to molar volume of molecules

DEMONSTRATIONS

1. Properties of gases: J. Chem. Educ. 60 67 (1983)
2. Boyle's law: J. Chem. Educ. 56 322 (1979); Shak. 2 12, 20
3. Charles' law: J. Chem. Educ. 56 823 (1979); 64 969 (1987); Shak. 2 24, 28
4. Determination of molar mass of butane: Shak. 2 48
5. Combining volumes of gases: Shakhashiri Videotapes, Demonstration 30
6. Dalton's law: Test. Dem. 156, 195; Shak. 2 41
7. Kinetic molecular theory simulator: Shak. 2 96

*8. Graham's law: Shak. 2 69, 73

9. Effusion of gases: Test. Dem. 9, 64, 128, 160, 188, 204; Shak. 2 59

* See also Shakhashiri Videotapes, Demonstration 32

QUIZZES

Quiz 1

1. Consider the reaction of aluminum with acid:

   $$2Al(s) + 6H^+(aq) \longrightarrow 2Al^{3+}(aq) + 3H_2(g)$$

   What mass of Al (molar mass = 26.98 g/mol) is required to produce 1.00 L of hydrogen at 740 mm Hg and 15°C?

2. A gas mixture at a total pressure of 842 mm Hg contains 0.100 mol of nitrogen and 0.800 mol of neon. What are the partial pressures of the two gases?

Quiz 2

1. Calculate the density of $SO_2(g)$ at 22°C and 734 mm Hg.
2. At a certain temperature and pressure it takes 52 s for 0.100 mol of H_2 to effuse through a pinhole. How long will it take for the same amount of Cl_2 to effuse at the same temperature and pressure?

Quiz 3

1. How many moles of nitrogen are there in a 20.0 L cylinder at

25°C if the pressure is 2190 lb/in^2? (14.7 lb/in^2 = 1 atm)

2. A sample of a certain gas weighing 2.51 g occupies a volume of 1.00 L at STP. What will be the volume of 1.00 g of this gas at 25°C and 1.00 atm?

Quiz 4

1. What volume of oxygen gas, at 751 mm Hg and 23°C, is formed by the decomposition of 6.48 g of $KClO_3$?

$$2KClO_3(s) \longrightarrow 2KCl(s) + 3\ O_2(g)$$

2. How many moles of hydrogen are there in a sample collected over water if the sample volume is 225 mL at 15°C and a total pressure of 740 mm Hg? (v. p. water at 15°C = 13 mm Hg)

Quiz 5

1. What is the molar mass of a gas which has a density of 0.00249 g/mL at 20°C and 744 mm Hg?

2. For an ideal gas, sketch a graph of:

a. P vs T   u vs P (constant T)

Answers

Quiz 1 1. 0.741 g 2. P_{N_2} = 93.6 mm Hg; P_{He} = 748 mm Hg

Quiz 2 1. 2.55 g/L 2. 310 s

Quiz 3 1. 122 mol 2. 0.434 L

Quiz 4 1. 1.95 L 2. 0.00910 mol

Quiz 5 1. 61.2 g/mol
2. a. straight line through origin
b. straight line parallel to x axis

PROBLEMS

1. $V = 11\ ft \times 12\ ft \times 8.0\ ft \times \frac{28.32\ L}{1\ ft^3} = 3.0 \times 10^4\ L$

$n = 3.541 \times 10^4\ g \times \frac{1\ mol}{29.0\ g} = 1.22 \times 10^3\ mol$

$T_K = 273 + 25 = 298\ K$

3. 1 atm = 760 mm Hg = 1.01325 kPa

728 mm Hg = 0.958 atm = 97.0 kPa

973 mm Hg = 1.28 atm = 1.30×10^2 kPa

749 mm Hg = 0.985 atm = 99.8 kPa

5. a. $5.75 \text{ L} \times \dfrac{0.890 \text{ atm}}{1.25 \text{ atm}} = 4.09 \text{ L}$

b. $5.75 \text{ L} \times \dfrac{0.890 \text{ atm}}{0.350 \text{ atm}} = 14.6 \text{ L}$

7. a. $1.75 \text{ L} \times \dfrac{333 \text{ K}}{303 \text{ K}} = 1.92 \text{ L}$

b. $1.75 \text{ L} \times \dfrac{288 \text{ K}}{303 \text{ K}} = 1.66 \text{ L}$

9. $P = 22.7 \text{ lb/in}^2 \times \dfrac{268 \text{ K}}{293 \text{ K}} = 20.8 \text{ lb/in}^2$

gauge $P = 20.8 \text{ lb/in}^2 - 14.7 \text{ lb/in}^2 = 6.1 \text{ lb/in}^2$

11. $V = 2.90 \text{ cm}^3 \times \dfrac{1.98 \text{ atm}}{1.50 \text{ atm}} \times \dfrac{288 \text{ K}}{281 \text{ K}} = 3.92 \text{ cm}^3$

13. $T_1 = 293 \text{ K}$; $P_1 = 0.980 \text{ atm}$

$n_2 = 0.200 \text{ mol} \times \dfrac{1.10 \text{ atm}}{0.980 \text{ atm}} \times \dfrac{293 \text{ K}}{306 \text{ K}} = 0.215 \text{ mol}$

15. $P_1 = 65 \text{ lb/in}^2$; $P_2 = 35 \text{ lb/in}^2$; $V_1 = 1.50 \text{ m}^3 \times 0.17 = 0.26 \text{ m}^3$

$V_2 = 0.26 \text{ m}^3 \times \dfrac{65 \text{ lb/in}^2}{35 \text{ lb/in}^2} = 0.48 \text{ m}^3$

$V = 0.48 \text{ m}^3 - 0.26 \text{ m}^3 = 0.22 \text{ m}^3$

17. $V = \dfrac{nRT}{P} = \dfrac{(47.8/4.003 \text{ mol})(0.0821 \text{ L}\cdot\text{atm/mol}\cdot\text{K})(306 \text{ K})}{2.25 \text{ atm}} = 133 \text{ L}$

19. $n = \dfrac{PV}{RT} = \dfrac{(1.00 \text{ atm})(162 \text{ L})}{(0.0821 \text{ L}\cdot\text{atm/mol}\cdot\text{K})(373 \text{ K})} = 5.29 \text{ mol}$

no. molecules $= 5.29 \text{ mol} \times \dfrac{6.022 \times 10^{23} \text{ molecules}}{1 \text{ mol}}$

$= 3.19 \times 10^{24}$ molecules

$V = 5.29 \text{ mol} \times \dfrac{18.02 \text{ g}}{1 \text{ mol}} \times \dfrac{1 \text{ cm}^3}{1.00 \text{ g}} = 95.3 \text{ cm}^3$

21. $\mathcal{M} = 17.03$ g/mol

P	V	T	n	m
2.50 atm	16.9 L	0°C	1.88 mol	32.0 g
7.50 atm	75.0 mL	30°C	0.0226 mol	0.385 g
768 mm Hg	6.0 L	100°C	0.20 mol	3.4 g
195 kPa	58.7 L	913°C	1.16 mol	19.8 g

23. $d = P\mathcal{M}/RT$

a. $d = \dfrac{(763/760 \text{ atm})(352.0 \text{ g/mol})}{(0.0821 \text{ L·atm/mol·K})(400 \text{ K})} = 10.8 \text{ g/L}$

b. $d = \dfrac{(763/760 \text{ atm})(28.01 \text{ g/mol})}{(0.0821 \text{ L·atm/mol·K})(400 \text{ K})} = 0.856 \text{ g/L}$

c. $d = \dfrac{(763/760 \text{ atm})(70.90 \text{ g/mol})}{(0.0821 \text{ L·atm/mol·K})(400 \text{ K})} = 2.17 \text{ g/L}$

25. $d\ H_2O(g) = \dfrac{(1.00 \text{ atm})(18.02 \text{ g/mol})}{(0.0821 \text{ L·atm/mol·K})(373 \text{ K})} = 0.588 \text{ g/L}$

$\text{ratio} = \dfrac{0.958 \text{ g/cm}^3}{0.000588 \text{ g/cm}^3} = 1.63 \times 10^3$

27. a. $\mathcal{M} = dRT/P = \dfrac{1.71 \text{ g/L} \times 0.0821 \text{ L·atm/mol·K} \times 298 \text{ K}}{755/760 \text{ atm}}$

$= 42.1 \text{ g/mol}$

b. $n\ C = 0.857 \times 42.1 \text{ g} \times \dfrac{1 \text{ mol}}{12.01 \text{ g}} = 3.00 \text{ mol}$

$n\ H = 0.143 \times 42.1 \text{ g} \times \dfrac{1 \text{ mol}}{1.008 \text{ g}} = 6.00 \text{ mol}$ $\quad C_3H_6$

29. a. $0.100(32.00 \text{ g/mol}) + 0.100(28.01 \text{ g/mol}) + 0.800(4.003 \text{ g/mol})$

$= 9.20 \text{ g/mol}$

b. $9.20/32.00 = 0.288$

31. $\mathcal{M} = \dfrac{mRT}{PV} = \dfrac{(2.00 \text{ g})(0.0821 \text{ L·atm/mol·K})(293 \text{ K})}{(1.00 \text{ atm})(0.3295 \text{ L})} = 146 \text{ g/mol}$

$32.07 + 6x = 146;\ x = 19;\ X = F$

33. a. $2H_2S(g) + 3\ O_2(g) \longrightarrow 2SO_2(g) + 2H_2O(g)$

b. $4.0 \text{ L } O_2 \times \dfrac{2 \text{ L } SO_2}{3 \text{ L } O_2} = 2.7 \text{ L } SO_2$

35. $n\ HCN = \dfrac{PV}{RT} = \dfrac{(751/760 \text{ atm})(8.53 \text{ L})}{(0.0821 \text{ L·atm/mol·K})(295 \text{ K})} = 0.348 \text{ mol}$

$m = 0.348 \text{ mol HCN} \times \dfrac{1 \text{ mol NaCN}}{1 \text{ mol HCN}} \times \dfrac{49.01 \text{ g}}{1 \text{ mol}} = 17.1 \text{ g}$

37. a. $n\ H_2O_2 = 5.50 \text{ mL} \times \dfrac{1.01 \text{ g}}{1 \text{ mL}} \times 0.0300 \times \dfrac{1 \text{ mol}}{34.02 \text{ g}} = 0.00490 \text{ mol}$

$$n\ O_2 = 0.00490\ mol\ H_2O_2 \times \frac{1\ mol\ O_2}{2\ mol\ H_2O_2} = 0.00245\ mol\ O_2$$

$$V = \frac{(0.00245\ mol)(0.0821\ L\cdot atm/mol\cdot K)(300\ K)}{745/760\ atm} = 0.0616\ L$$

b. $5.50\ mL \times 10 = 55.0\ mL$; yes

39. $P\ N_2 = 0.745 \times 751\ mm\ Hg = 559\ mm\ Hg$

$P\ O_2 = 118\ mm\ Hg$; $P\ CO_2 = 27\ mm\ Hg$; $P\ H_2O = 47\ mm\ Hg$

41. a. $P\ O_2 = 775\ mm\ Hg - 26.7\ mm\ Hg = 748\ mm\ Hg$

b. $m = \frac{PV\mathcal{M}}{RT} = \frac{(748/760\ atm)(0.355\ L)(32.00\ g/mol)}{(0.0821\ L\cdot atm/mol\cdot K)(300\ K)} = 0.454\ g$

43. $$n\ O_2 = \frac{(721/760\ atm)(12.83\ L)}{(0.0821\ L\cdot atm/mol\cdot K)(298\ K)} = 0.497\ mol$$

$$V = \frac{(0.497\ mol)(0.0821\ L\cdot atm/mol\cdot K)(323K)}{762/760\ atm} = 13.1\ L$$

45. a. lighter b. $1.25 = (\mathcal{M}NF_3/\mathcal{M})^{\frac{1}{2}}$; ratio $= (1.25)^2 = 1.56$

47. $r\ He/r\ Ne = (20.18/4.003)^{\frac{1}{2}} = 2.245$

$\Delta P = 2.245 \times 11\ mm\ Hg = 25\ mm\ Hg$

$P = 760\ mm\ Hg - 25\ mm\ Hg = 735\ mm\ Hg$

49. a. $u_2/u_1 = 1/2 = (T_2/298)^{\frac{1}{2}}$; $T_2 = 74.5\ K$ b. no

51. a. $u/482\ m/s = (32.00/70.90)^{\frac{1}{2}} = 0.672$; $324\ m/s$

b. $u/482\ m/s = (248 \times 32.00/298 \times 39.95)^{\frac{1}{2}} = 0.816$; $394\ m/s$

53. a. CO b. increase; decrease

55. $V_m/V^\circ_m \approx 0.78$; $V^\circ_m = \frac{(0.0821\ L\cdot atm/mol\cdot K)(298\ K)}{200\ atm} = 0.122\ L$

$V_m \approx 0.78 \times 0.122\ L = 0.095\ L$

$d = 16.04\ g/0.095\ L = 1.70 \times 10^2\ g/L$

$d^\circ = 16.04\ g/0.122\ L = 1.31 \times 10^2\ g/L$

57. p. 123

59. $2CH_3OH(l) + 3\ O_2(g) \longrightarrow 2CO_2(g) + 4H_2O(l)$

$$10\ mi \times \frac{1\ gal}{32\ mi} \times \frac{4\ qt}{1\ gal} \times \frac{1\ L}{1.057\ qt} \times \frac{791\ g}{1\ L} = 940\ g$$

$$940 \text{ g } CH_3OH \times \frac{1 \text{ mol } CH_3OH}{32.04 \text{ g } CH_3OH} \times \frac{1 \text{ mol } CO_2}{1 \text{ mol } CH_3OH} = 29 \text{ mol } CO_2$$

$$V = \frac{(29 \text{ mol})(0.0821 \text{ L·atm/mol·K})(298 \text{ K})}{1.00 \text{ atm}} = 7.1 \times 10^2 \text{ L}$$

61. $d_1h_1 = d_2h_2$; $0.789 \frac{g}{mL} \times h_1 = 13.6 \frac{g}{mL} \times 760$ mm

$h_1 = 1.31 \times 10^4$ mm (about 43 ft)

63. a. straight line through origin
b. straight line through origin
c. hyperbola
d. straight line through origin

65. a. 1.00 b. 38.00/28.02 = 1.356 c. 1.00 d. 1.00

67. $$1 \text{ gal} \times \frac{4 \text{ qt}}{1 \text{ gal}} \times \frac{1 \text{ L}}{1.057 \text{ qt}} \times \frac{692 \text{ g}}{1 \text{ L}} = 2.62 \times 10^3 \text{ g octane}$$

$$n\ O_2 = 2.62 \times 10^3 \text{ g } C_8H_{18} \times \frac{1 \text{ mol } C_8H_{18}}{114.22 \text{ mol } C_8H_{18}} \times \frac{25 \text{ mol } O_2}{2 \text{ mol } C_8H_{18}}$$

$$= 287 \text{ mol } O_2$$

$$V\ O_2 = \frac{(287 \text{ mol})(0.0821 \text{ L·atm/mol·K})(298 \text{ K})}{1.00 \text{ atm}} = 7.01 \times 10^3 \text{ L}$$

V air = 7.01×10^3 L/0.210 = 3.34×10^4 L

69. mass CO_2 (ideal gas law) = 0.239 g

mass C = 0.0652 g; mass H = 0.0136 g; 32.0% C; 6.68% H

mass N_2 (ideal gas law) = 0.0467 g; 18.7% N; 42.6% O

empirical formula from percentages: $C_2H_5O_2N$

71. a. n_I = 1000/44.01 = 22.72

n_{II} = 1000/28.02 = 35.69; cylinder II

b. $n_I = \frac{7.0 \text{ V}}{323 \text{ R}} = 0.0217$ V/R ; cylinder I

$n_{II} = \frac{5.5 \text{ V}}{283 \text{ R}} = 0.0194$ V/R

72. r NH_3/r HCl = d NH_3/d HCl = $(36.45/17.04)^{\frac{1}{2}}$ = 1.463

2.463 d HCl = 3.0 ft; d HCl = 1.2 ft

1.8 ft from NH_3 end

73. $1.8°R = K$

$$0.0821\ \frac{L\cdot atm}{mol\cdot K} \times \frac{1\ K}{1.8°R} = 0.0456\ \frac{L\cdot atm}{mol\cdot °R}$$

74. Let x = mass Al

$$n\ H_2 = \frac{(0.153\ L)(1.00\ atm)}{(0.0821\ L\cdot atm/mol\cdot K)(298\ K)} = 0.00625\ mol$$

$$0.00625 = \frac{3x/2}{26.98} + \frac{(0.2500 - x)}{65.39}\ ;\ x = 0.0603$$

$$\%\ Al = \frac{0.0603}{0.2500} \times 100 = 24.1$$

75. $V = 4\pi r^3/3$

$$m_{air} = \frac{(752/760\ atm)(V)(29.0\ g/mol)}{(0.0821\ L\cdot atm/mol\cdot K)(295\ K)} = 1.18\ V$$

$$m\ H_2 = \frac{(752/760\ atm)(V)(2.016\ g/mol)}{(0.0821\ L\cdot atm/mol\cdot K)(295\ K)} = 0.0824\ V$$

$$1.18\ V = 1.75 \times 10^5 + 0.0824\ V;\ V = 1.59 \times 10^5\ L = 159\ m^3$$

$$159\ m^3 = 4\pi r^3/3;\ r = 3.36\ m;\ diameter = 6.72\ m$$

76. $n\ H_2O = 1.50;\ n\ H_2 = 0.50;\ n\ O_2 = 0.25;\ n\ tot = 2.25$

$$P_2 = P_1 \times \frac{n_2}{n_1} \times \frac{T_2}{T_1} = 0.820\ atm \times \frac{2.25}{3.00} \times \frac{398\ K}{298\ K} = 0.821\ atm$$

77. $V_a = n_aRT/P;\quad V_{tot} = n_{tot}RT/P$

$$V_a/V = n_a/n_{tot} = X_a$$

$n_a/n_{tot} \neq m_a/m_{tot}$ because molar masses differ

CHAPTER 6
Electronic Structure and the Periodic Table

LECTURE NOTES

This chapter requires about 3 lectures. Experience suggests that all four quantum numbers are difficult for students to digest at a single sitting. We prefer to break them up between two lectures.

The "aufbau" approach can be used to predict electronic structure of atoms through atomic number 36. Beyond that, it's probably simpler to use the periodic table to derive electronic structures. Orbital diagrams follow from electron configurations by applying Hund's rule.

In discussing the electronic structure of monatomic ions, it is important to stress the difference between transition metal atoms and cations. The chapter concludes with a description of trends in the periodic table as regards atomic properties.

LECTURE 1

I Atomic Spectra

Produced when electron moves from higher to lower energy level, giving off light in the process.

$$\Delta E = E_{hi} - E_{lo} = h\nu = hc/\lambda$$

For the yellow line in the sodium spectrum, $\lambda = 589.0$ nm

$$\nu = \frac{c}{\lambda} = \frac{2.998 \times 10^{8} \text{ m/s}}{589 \times 10^{-9} \text{ m}} = 5.090 \times 10^{14}/\text{s}$$

$$\Delta E = \frac{(6.626 \times 10^{-34} \text{ J}\cdot\text{s})(2.998 \times 10^{8} \text{ m/s})}{589.0 \times 10^{-9} \text{ m}} = 3.373 \times 10^{-19} \text{ J}$$

For one mole of electrons:

$$\Delta E = 3.373 \times 10^{-19} \text{ J} \times \frac{6.022 \times 10^{23}}{1 \text{ mol}} \times \frac{1 \text{ kJ}}{10^{3} \text{ J}} = 203.1 \text{ kJ}$$

Hence two energy levels in the Na atom differ in energy by 203.1 kJ/mol.

II Hydrogen Atom

A. Bohr model   Bohr postulated that electron moves about nucleus in circular orbit of fixed radius. By absorbing energy, it moves to a higher orbit of larger energy; energy given off as electron returns.

$E_n = (-2.180 \times 10^{-18}\ J)/n^2 \qquad n = 1, 2, 3, ---$

When electron moves from n = 3 to n = 2:

$E_3 = -2.422 \times 10^{-19}\ J;\ E_2 = -5.450 \times 10^{-19}\ J$

$E_{hi} - E_{lo} = 3.028 \times 10^{-19}\ J$

$$\lambda = \frac{hc}{\Delta E} = \frac{(6.626 \times 10^{-34}\ J\cdot s)(2.998 \times 10^{8}\ m/s)}{3.028 \times 10^{-19}\ J}$$

$$= 6.560 \times 10^{-7}\ m = 656.0\ nm$$

(1st line in Balmer series)

B. Quantum mechanical model
1. Can only refer to the probability of finding an electron in a region; cannot specify path.
2. Four quantum numbers required to describe energy of electron completely in all atoms.

III Electronic Structure

A. Principal energy levels
n = 1, 2, 3, - - . Value of n is the main factor that determines the energy of an electron and its distance from the nucleus. Maximum capacity of principal level = $2n^2$.

n	1	2	3	4
max. no. e^-	2	8	18	32

B. Sublevels
1. Quantum number ℓ = 0, 1, 2, - - (n - 1)

n = 1 $\quad \ell$ = 0 (one sublevel)
n = 2 $\quad \ell$ = 0, 1 (two sublevels)
n = 3 $\quad \ell$ = 0, 1, 2 (three sublevels)

In general, no. of sublevels = n

2. Sublevel designations: s, p, d, f

value of ℓ	0	1	2	3
letter	s	p	d	f
capacity	2	6	10	14

LECTURE 2

I Electronic Structure

A. Electron configuration Indicate by a superscript the number of electrons in each sublevel.

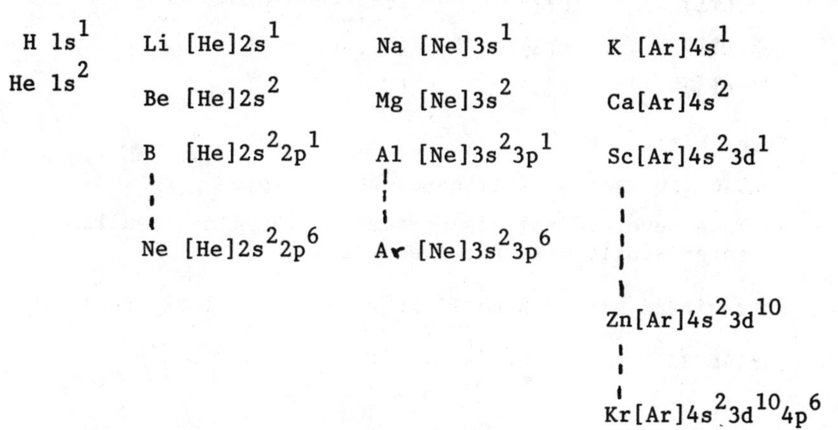

H $1s^1$	Li $[He]2s^1$	Na $[Ne]3s^1$	K $[Ar]4s^1$
He $1s^2$	Be $[He]2s^2$	Mg $[Ne]3s^2$	Ca$[Ar]4s^2$
	B $[He]2s^22p^1$	Al $[Ne]3s^23p^1$	Sc$[Ar]4s^23d^1$
	⋮	⋮	⋮
	Ne $[He]2s^22p^6$	Ar $[Ne]3s^23p^6$	⋮
			Zn$[Ar]4s^23d^{10}$
			⋮
			Kr$[Ar]4s^23d^{10}4p^6$

Beyond krypton, it's simplest to derive electron configurations from the periodic table.

Groups 1, 2: fill s sublevel
Groups 13-18: fill p sublevel
Groups 3-12: fill d sublevels (transition metals)
Lanthanides and actinides fill f sublevels (4f, 5f)

Electron configuration iodine atom? $[Kr]5s^24d^{10}5p^5$

B. 3rd and 4th quantum numbers

1. Orbital designated by $m_\ell = \ell, ---, 0, ---, -\ell$

   $\ell = 0$ (s sublevel); $m_\ell = 0$ (one s orbital)
   $\ell = 1$ (p sublevel); $m_\ell = 1, 0, -1$ (three p orbitals)
   $\ell = 2$ (d sublevel); $m_\ell = 2, 1, 0, -1, -2$ (5 d orbitals)

   Each orbital has a capacity of two electrons. s orbitals are spherically symmetric about nucleus; p orbitals are dumbell shaped and are at right angles to each other.

2. Electron in an orbital can have either of two spins:

   $m_s = +\frac{1}{2}, -\frac{1}{2}$

C. Orbital diagram Show number of electrons in each orbital and spin of each electron.

	1s
H	(↑)
He	(↑↓)

	1s	2s	2p
Li	(↑↓)	(↑)	
Be	(↑↓)	(↑↓)	
B	(↑↓)	(↑↓)	(↑) ( ) ( )
C	(↑↓)	(↑↓)	(↑) (↑) ( )
N	(↑↓)	(↑↓)	(↑) (↑) (↑)
O	(↑↓)	(↑↓)	(↑↓) (↑) (↑)

Note that:

1. 2 e^- in same orbital have opposed spins
2. When several orbitals of same sublevel are available, e^- enter singly with parallel spins.

Abbreviated electron configuration, orbital diagram of Fe?

$[Ar]4s^2 3d^6$

	4s	3d
[Ar]	(↑↓)	(↑↓) (↑) (↑) (↑) (↑)

LECTURE 3

I Electronic Structure

A. Monatomic ions

1. Ions with noble gas structures (Groups 1, 2, 16, 17)
2. Transition metal cations; outer s electrons are lost

$_{24}Cr^{3+}$ $[Ar]3d^3$ $_{27}Co^{2+}$ $[Ar]3d^7$ $_{30}Zn^{2+}$ $[Ar]3d^{10}$

II Trends in Periodic Table

A. Atomic radius

1. In general, atomic radius decreases going across a period from left to right, increases going down group.

Na 0.186 nm	Mg 0.160 nm	S 0.104 nm	Cl 0.099 nm
K 0.231 nm			Br 0.114 nm

2. Trends can be explained in terms of effective nuclear charge felt by outer electron(s). Eff. nucl. charge = Z - S, where Z is nuclear charge, S is number of electrons in inner, complete levels. Electrons in outer levels do not shield one another effectively

Na: 11 - 10 = 1 Mg: 12 - 10 = 2 Al: 13 - 10= 3
K: 19 - 18 = 1

B. Ionic radius

Trends parallel those in atomic radius. Beyond

that:

cations are smaller than the corresponding atoms
anions are larger than the corresponding atoms

This means that, in a typical ionic compound, the anions occupy most of the space.

C. Ionization energy = energy that must be absorbed to convert an atom to a +1 ion

$Na(g) \longrightarrow Na^+(g) + e^-$ I.E. = +496 kJ/mol

increases → in periodic table, as atoms get smaller
decreases ↓ in periodic table, as atoms get larger

D. Electronegativity Measure of attraction of atom for electrons in covalent bond.

increases →, decreases ↓

DEMONSTRATIONS

1. Flame tests: Test. Dem. 22, 91; J. Chem. Educ. 65 452, 544 (1988), 67 791 (1990), 68 937 (1991), 69 327 (1992)

*2, Singlet molecular oxygen: Shak. 1 133

3. Paramagnetism: Test. Dem. 215

* See also Shakhashiri Videotapes, Demonstration 33

QUIZZES

Quiz 1

1. Consider the germanium atom (Z = 32)
   a. Write the electron configuration of the Ge atom.
   b. Write the abbreviated orbital diagram of Ge.
   c. Compare Ge to C in atomic radius, ionization energy, and electronegativity.

Quiz 2

1. Calculate the wavelength of the line in the Lyman series of hydrogen resulting from the transition: n = 3 to n = 1

$E = \frac{-2.180 \times 10^{-18}\ J}{n^2}$; $h = 6.626 \times 10^{-34}$ J•s
$c = 2.998 \times 10^{10}$ m/s

2. Write the abbreviated electron configuration of the Mn^{2+} ion.

Quiz 3

1. A certain spectral line originates from an electron transition between two levels that differ in energy by 282 kJ/mol. Calculate the wavelength and frequency of the line.

   $h = 6.626 \times 10^{-34}$ J·s; $c = 2.998 \times 10^{10}$ m/s

   $N_A = 6.022 \times 10^{23}$/mol

2. Write the abbreviated electron configuration of the Ni atom (at. no. = 28). How many unpaired electrons are there in this atom?

Quiz 4

1. Sketch the geometry of an s orbital; of the three p orbitals.
2. Give the orbital diagram of the Co^{2+} ion.

Quiz 5

1. Explain, in terms of effective nuclear charge, why atomic radius decreases moving across the periodic table from left to right.
2. Give the electron configuration of the atom and the +2 ion of Ti (Z = 22).

Answers

Quiz 1 a. $1s^2 2s^2 2p^6 3s^2 3p^6 4s^2 3d^{10} 4p^2$

b. [Ar] 4s (↑↓) 3d (↑↓)(↑↓)(↑↓)(↑↓)(↑↓) 4p (↑)(↑)( )

c. Ge has larger atomic radius, smaller ionization energy, is less electronegative

Quiz 2 1. 102.6 nm 2. $[Ar]3d^5$

Quiz 3 1. 424 nm, 7.07×10^{14}/s 2. $[Ar]4s^2 3d^8$; 2

Quiz 4 1. s orbital is spherical; p orbitals are dumbell-shaped at right angles to each other.

2. 1s (↑↓) 2s (↑↓) 2p (↑↓)(↑↓)(↑↓) 3s (↑↓) 3p (↑↓)(↑↓)(↑↓)

3d (↑↓)(↑↓)(↑)(↑)(↑)

Quiz 5 1. Effective nuclear charge increases moving across table because outer electrons are ineffective in screening. Nuclear charge increases, number of inner electrons remains constant.

2. $1s^2 2s^2 2p^6 3s^2 3p^6 4s^2 3d^2$; $1s^2 2s^2 2p^6 3s^2 3p^6 3d^2$

PROBLEMS

1. Photon with short wavelength has higher frequency, larger energy.

3. a. $\nu = \frac{c}{\lambda} = \frac{2.998 \times 10^{8} \text{ m/s}}{585 \times 10^{-9} \text{ m}} = 5.12 \times 10^{14}/\text{s}$

   b. $E = h\nu = (6.626 \times 10^{-34} \text{ J·s})(5.12 \times 10^{14}/\text{s})$
   $= 3.39 \times 10^{-19} \text{ J}$

   c. $E = 3.39 \times 10^{-19} \text{ J} \times \frac{1 \text{ kJ}}{10^3 \text{ J}} \times \frac{6.022 \times 10^{23}}{\text{mol}} = 204 \text{ kJ/mol}$

5. $E = \frac{(6.626 \times 10^{-34} \text{ J·s})(2.998 \times 10^{8} \text{ m/s})}{80 \times 10^{-9} \text{ m}} \times \frac{1 \text{ kJ}}{10^3 \text{ J}} \times \frac{6.022 \times 10^{23}}{\text{mol}}$
   $= 1.50 \times 10^{3}$ kJ/mol; yes

7. $\Delta E = 5.00 \times 10^{2} \frac{\text{kJ}}{\text{mol}} \times \frac{10^3 \text{ J}}{1 \text{ kJ}} \times \frac{1 \text{ mol}}{6.022 \times 10^{23}} = 8.30 \times 10^{-19} \text{ J}$

   a. $\lambda = \frac{(6.626 \times 10^{-34} \text{ J·s})(2.998 \times 10^{8} \text{ m/s})}{8.30 \times 10^{-19} \text{ J}}$
   $= 2.39 \times 10^{-7} \text{ m} = 239 \text{ nm}$

   b. $\nu = \frac{2.998 \times 10^{8} \text{ m/s}}{2.39 \times 10^{-7} \text{ m}} = 1.25 \times 10^{15}/\text{s}$

9. a. $\lambda = \frac{2.998 \times 10^{8} \text{ m/s}}{6.6 \times 10^{14}/\text{s}} = 4.5 \times 10^{-7} \text{ m} = 4.5 \times 10^{2} \text{ nm}$

   b. $E = h\nu = (6.626 \times 10^{-34} \text{ J·s})(6.6 \times 10^{14}/\text{s}) = 4.4 \times 10^{-19} \text{ J}$

11. a. (1) b. (2) c. (3)

13. Concentric circles of steadily increasing radius
    a. transition from higher levels to n = 1
    b. transition from higher levels to n = 2

15. $\Delta E = 2.180 \times 10^{-18} \text{ J}(1/16 - 1/36) = 7.569 \times 10^{-20} \text{ J}$

    $\lambda = \frac{(6.626 \times 10^{-34} \text{ J·s})(2.998 \times 10^{8} \text{ m/s})}{7.569 \times 10^{-20} \text{ J}} = 2.624 \times 10^{-6} \text{ m}$
    $= 2.624 \times 10^{3} \text{ nm}$

17. $\Delta E = \frac{(6.626 \times 10^{-34}\ J\cdot s)(2.998 \times 10^{8}\ m/s)}{434.05 \times 10^{-9}\ m} = 4.577 \times 10^{-19}\ J$

$4.577 \times 10^{-19} = 2.180 \times 10^{-18}(1/4 - 1/n^2)$

$n^2 = 25.0$; n = 5

19. Quantum mechanical model does not locate electron precisely, requires four quantum numbers to describe energy completely.

21. a. 2, 1, 0, -1, -2   b. 4, 3, 2, 1, 0, -1, -2, -3, -4

c. ℓ = 3: 3, 2, 1, 0, -1, -2, -3
ℓ = 2: 2, 1, 0, -1, -2
ℓ = 1: 1, 0, -1
ℓ = 0: 0

23. a. 4p   b. 4d   c. 3d   d. 4f

25. a. p   b. f   c. s

27. a. 5 d orbitals + 3 p orbitals + 1 s orbital = 9
b. 5   c. 7

29. a. ℓ = 0, 1, - - (n - 1)
b. $m_\ell = \ell, \ell - 1$, - - , 0, -1, - - , $-\ell$
c. none

31. a. no 1p orbital   c. m_ℓ cannot equal 1 when ℓ = 0
d. m_s cannot equal 0

33. a. $1s^22s^22p^63s^23p^5$

b. $1s^22s^22p^63s^23p^64s^13d^5$

c. $1s^22s^22p^63s^23p^64s^23d^{10}4p^4$

d. $1s^22s^22p^63s^23p^2$

e. $1s^22s^22p^63s^23p^64s^23d^{10}4p^65s^24d^{10}5p^4$

35. a. $[Ar]4s^23d^{10}4p^5$ b. $[Ar]4s^23d^2$ c. $[Kr]5s^24d^{10}$

d. $[He]2s^22p^5$ e. $[Xe]6s^24f^{14}5d^{10}$

37. a. Sc   b. Pr   c. He   d. Ne

39. a. 6/18 = 0.333   b. 10/48 = 0.208   c. 4/10 = 0.400

41. a. impossible   b. excited   c. ground   d. excited

e. impossible   f. excited

43. a. 1s (↑↓)   2s (↑↓)   2p (↑)(↑)( )

b. 1s (↑↓)   2s (↑↓)   2p (↑↓)(↑↓)(↑↓)   3s (↑↓)   3p (↑↓)(↑↓)(↑↓)   4s (↑↓)

3d (↑↓)(↑)(↑)(↑)(↑)

c. 1s (↑↓)   2s (↑↓)   2p (↑↓)(↑↓)(↑↓)   3s (↑↓)   3p (↑)(↑)(↑)

d. (↑↓)   (↑↓)   (↑↓)(↑↓)(↑↓)   (↑↓)   (↑↓)(↑↓)(↑↓)

45. a. Ni   b. Co   c. Ge

47. a. Ar and all succeeding elements

b. Ca, Zn   c. Cl   d. none

49. a. 2   b. 0   c. 3

51. a. Zn   b. Sc, Cu   c. Ti, Ni   d. V, Co   e. Fe   f. Mn

53. a. $1s^2 2s^2 2p^6 3s^2 3p^6 4s^2$; $1s^2 2s^2 2p^6 3s^2 3p^6$

b. $1s^2 2s^2 2p^6 3s^2 3p^6 4s^2 3d^{10} 4p^4$; $1s^2 2s^2 2p^6 3s^2 3p^6 4s^2 3d^{10} 4p^6$

c. $1s^2 2s^2 2p^6 3s^2 3p^6 4s^2 3d^{10} 4p^6 5s^2 4d^2$

$1s^2 2s^2 2p^6 3s^2 3p^6 4s^2 3d^{10} 4p^6 4d^2$

d $1s^2 2s^2 2p^6 3s^2 3p^6 3d^7$; $1s^2 2s^2 2p^6 3s^2 3p^6 3d^6$

55. a. 3   b. 4   c. 0   d. 0

57. a. I < Te < Rb   b. Rb < Te < I   c. Rb < Te < I

59. a. S   b. K   c. K

61. a. K   b. O^{2-} c. Tl   d. Ni^{2+}

63. a. Rb > K > Ca > Ca^{2+} b. Te^{2-} > Te > Se > S

65. $[Xe]6s^2 4f^5$; Np

67. $[Rn]7s^2 5f^{14}$; Yb

69. E light = 75 J x 0.07 = 5.25 J

$$E\ \text{photon} = \frac{(6.626 \times 10^{-34}\ \text{J}\cdot\text{s})(2.998 \times 10^{8}\ \text{m/s})}{565 \times 10^{-9}\ \text{m}}$$

$$= 3.52 \times 10^{-19}\ \text{J}$$

no. photons = $5.25/(3.52 \times 10^{-19}) = 1.5 \times 10^{19}$

71. a. Bohr model specifies position of electron; quantum mechanical model deals with probability.

    b. See Figure 6.1

    c. attraction into magnetic field vs slight repulsion

    d. at 90° angles to each other

73. a. false; evolves energy b. false; longer wavelength

    c. presumably true (2, 6, 10, 14, 18) d. true

75. 5; 4, 3, 2, 1, 0, -1, -2, -3, -4 ; 18

77. a. F b. T c. F

79. Z = 3, n = 2

$$E = -2.180 \times 10^{-18}\ \text{J} \times \frac{9}{4} \times \frac{1\ \text{kJ}}{10^3\ \text{J}} \times \frac{6.022 \times 10^{23}}{\text{mol}}$$

$$= -2.954 \times 10^{3}\ \text{kJ}$$

$\Delta E = 0 - (-2954\ \text{kJ}) = 2954\ \text{kJ}$

80. $\Delta E = 2.180 \times 10^{-18}\ \text{J}\ (1/4 - 1/n^2)$

$$= 2.180 \times 10^{-18}\ \text{J}(n^2 - 4)/4n^2$$

$$\lambda = \frac{(6.626 \times 10^{-34}\ \text{J s})(2.998 \times 10^{8}\ \text{m/s})}{2.180 \times 10^{-18}(n^2 - 4)\ \text{J}} \times 4n^2 \times \frac{10^9\ \text{nm}}{1\ \text{m}}$$

$$= 3.645 \times 10^{2}\ n^2/(n^2 - 4)$$

81. n = 1; $\ell = 0$; $m_\ell = 0, 1$ $4e^-$

    $\ell = 1$; $m_\ell = 0, 1, 2$ $6e^-$

    n = 2; $\ell = 0$; $m_\ell = 0, 1$ $4e^-$

    $\ell = 1$; $m_\ell = 0, 1, 2$ $6e^-$

    $\ell = 2$; $m_\ell = 0, 1, 2, 3$ $8e^-$

    $1s^4 1p^4$

82. a. $3e^-$; $9e^-$; $15e^-$ b. $27e^-$

c. $1s^3 2s^3 2p^2$; $1s^3 2s^3 2p^9 3s^2$

83. a. 540 nm; $E = \frac{(6.626 \times 10^{-34}\ J{\cdot}s)(2.998 \times 10^8\ m/s)}{540 \times 10^{-9}\ m}$

$= 3.68 \times 10^{-19}\ J$

400 nm: $E = 4.97 \times 10^{-19}\ J$

$E_{min} = 3.68 \times 10^{-19}\ J - 0.26 \times 10^{-19}\ J = 3.42 \times 10^{-19}\ J$

b. with no kinetic energy:

$\lambda = \frac{(6.626 \times 10^{-34}\ J{\cdot}s)(2.998 \times 10^8\ m/s)}{3.42 \times 10^{-19}\ J}$

$= 5.81 \times 10^{-7}\ m = 581\ nm$

CHAPTER 7
Covalent Bonding

LECTURE NOTES

This chapter requires at least 2½, perhaps 3 lectures. The first lecture is devoted primarily to Lewis structures. The second lecture covers molecular geometry and polarity. The final lecture deals with hybridization, sigma and pi bonding. Note that

1. Students find Lewis structures relatively simple to draw provided they get sufficient practice; a molecular model kit helps. Note that the ability to write Lewis structures is essential to an understanding of just about everything else in the chapter, including molecular geometry and hybridization.

2. You may or may not wish to discuss formal charge (p. 170). It is useful in choosing between alternative skeletons, but is also somewhat of a diversion in a closely-packed chapter.

3. Table 7.3 is a useful way of summarizing the geometries of all species in which a central atom is surrounded by 2, 3, or 4 electron pairs. The AXE notation is useful, because it enables students to predict the geometries of multiple-bonded species as well.

4. Figure 7.7 shows the geometries of species where the central atom is surrounded by 5 or 6 electron pairs, including unshared pairs. In principle, these geometries can all be predicted by VSEPR theory; in practice, a certain amount of memorization is necessary.

5. Students often get the mistaken idea that promotion of electrons is necessary for hybridization to occur. They also have trouble distinguishing between sigma and pi bonds.

LECTURE 1

I Lewis Structures

A Lewis structure shows the distribution of outer (valence) electrons in an atom, molecule, or polyatomic ion. Unshared electrons are shown as dots, bonds as straight lines.

$$H\cdot + \cdot\ddot{\underset{..}{F}}: \longrightarrow H - \ddot{\underset{..}{F}}:$$

$$2H\cdot + \cdot\ddot{\underset{..}{O}}\cdot \longrightarrow H - \ddot{\underset{..}{O}} - H$$

In H_2O and HF, as in most molecules and polyatomic ions, nonmetal atoms, except H, are surrounded by 8 electrons, an octet. In this sense, each atom has a noble gas structure. Lewis structures are written following a stepwise procedure.

A. Rules for writing Lewis structures (single bonds)

1. Count valence electrons available. Number of valence electrons contributed by nonmetal atom is equal to the last digit of its group number in the periodic table (1 for H). Add electrons to take into account negative charge.

   OCl^- ion: 6 + 7 + 1 = 14 valence e^-

   CH_3OH molecule: 4 + 4(1) + 6 = 14 valence e^-

   SO_3^{2-} ion: 6 + 3(6) + 2 = 26 valence e^-

2. Draw skeleton structure, using single bonds

```
                   H
                   |
O - Cl       H  -  C  -  O  -  H        O - S - O
                   |                        |
                   H                        O
```

   Note that carbon almost always forms four bonds. Central atom is written first in formula. Terminal atoms are most often H, O, or a halogen.

3. Deduct two electrons for each single bond in the skeleton.

   OCl^- ion: 14 - 2 = 12 valence e^- left

   CH_3OH molecule: 14 - 10 = 4 valence e^- left

   SO_3^{2-} ion: 26 - 6 = 20 valence e^- left

4. Distribute these electrons to give each atom a noble gas structure, if possible.

```
                            H
   ..     ..                |               ..     ..     ..
( :O  -  Cl: )-       H  -  C  -  O  -  H   (:O  -  S  -  O:)2-
   ..     ..                |     ..          ..     |     ..
                            H                       :O:
                                                     ..
```

B. Too few electrons; form multiple bonds

   Structure of NO_3^- ion?

   no. of valence e^- = 5 + 18 + 1 = 24

   skeleton
```
O - N - O
    |
    O
```

   valence e^- left = 24 - 6 = 18

If all electrons are distributed as unshared pairs:

```
:Ö - N - Ö:
     |
    :Ö:
```

Here, N atom is surrounded by only 6 valence e^-. To remedy a deficiency of 2 electrons, form a double bond:

```
:Ö - N - Ö:
     ||
    :O:
```

N_2 structure? 10 valence e^- :N ≡ N:

C. <u>Too many electrons</u>: give central atom an expanded octet

Consider XeF_4: 36 valence e^- Octet structure uses 32 e^-

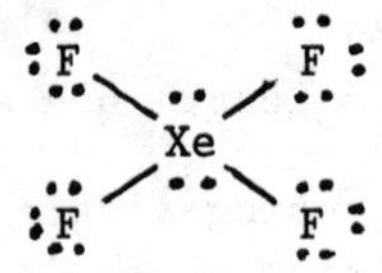

In a few molecules, there are less than 8 electrons around central atom.

```
:F̤ - Be - F̤:      :F̤ - B - F̤:
                       |
                      :F̤:
```

D. <u>Resonance forms</u>

To explain the fact that all three bonds in the nitrate ion are the same length, invoke the concept of resonance.

```
Ö = N - Ö:   ←→   :Ö - N - Ö:   ←→   :Ö - N = Ö
    |                  ||                  |
   :Ö:                :O:                 :Ö:
```

True structure is a "hybrid" of these three forms. Note that:

1. Resonance forms obtained by moving electrons, not atoms
2. Resonance can be expected when it is possible to draw more than one structure that follows the octet rule.

LECTURE 2

I <u>Molecular Geometry</u>

VSEPR principle: Electron pairs around a central atom tend to be oriented so as to be as far apart as possible.

A. <u>Two to six atoms around central atom</u>; no unshared pairs

BeF_2 linear, BF_3 triangular planar, CF_4 tetrahedral,

PF_5 triangular bipyramid, SF_6 octahedral. Discuss bond angle

B. <u>Unshared pairs</u> (Table 7.3)

AX_2E (GeF_2): bent, 120°

AX_3E (NH_3): triangular pyramid, 109°

AX_2E_2 (H_2O) : bent, 109°

Expanded octets: refer to Figure 7.7

C. Multiple bonding Has no effect upon geometry; Table 7.3 applies here as well.

BF_3 and SO_3: both AX_3 molecules, same geometry

BeF_2 and CO_2: both AX_2, both linear

II Polarity

A. Bond polarity All bonds are polar unless the two atoms joined are identical (H - H). Extent of polarity depends upon difference in electronegativity.

H - H	ΔE.N. = 0	nonpolar
H - C	ΔE.N. = 0.3	slightly polar
H - F	ΔE.N. = 1.8	strongly polar (- pole at F atom)

B. Molecular polarity

1. Diatomic molecules: polar if atoms differ

H - Cl	Cl - Cl
polar	nonpolar

HCl molecules line up in electric field, Cl_2 molecules don't

2. Polyatomic molecules Even though bonds are polar, molecule may be nonpolar if bonds are symmetrically distributed.

F← Be→ F	H ↗ O ↖ H	CH_4	CH_3Cl
nonpolar	polar	nonpolar	polar

LECTURE 2½

I Atomic Orbitals; Hybridization

In molecules, the orbitals occupied by electron pairs are seldom "pure" s or p orbitals. Instead, they are "hybrid" orbitals, formed by combining s, p, and d orbitals.

A. Formation of hybrid orbitals

1. s orbital + p orbital ⟶ two sp hybrid orbitals

	2s	2p
Be in BeF_2	(↑↓)	(↑↓)

2. s orbital + two p orbitals ⟶ three sp^2 hybrids

	2s	2p
B in BF_3	(↑↓)	(↑↓)(↑↓)

3. s orbital + three p orbitals ⟶ four sp^3 hybrids

C in CH_4 — 2s (↑↓) 2p (↑↓)(↑↓)(↑↓)

B. Hybridization with 5 or 6 electron pairs: sp^3d, sp^3d^2. Note that expanded octets do not occur with atoms in the second period (e.g., N, O, F) since there are no 2d orbitals.

C. Unshared pairs can be hybridized (H_2O, NH_3). Only one of the electron pairs in a multiple bond is hybridized.

:Ö - S̈ = Ö sp^2 hybridization for sulfur

Ö = C = Ö   sp hybridization for carbon

II Sigma and Pi Bonds

When a bond consists of an electron pair in a hybrid orbital, the electron density is concentrated along the bond axis and is symmetrical about it. Such a bond is called a sigma bond. The "extra" electron pairs in a multiple bond are located in unhybridized orbitals which are not concentrated along the bond axis. Such bonds are called pi bonds.

CO_2 two sigma bonds, two pi bonds

SO_2 two sigma bonds, one pi bond

N_2 one sigma bond, two pi bonds

DEMONSTRATIONS

1. Spontaneous detonation of odd-electron species: J. Chem. Educ. 68 938 (1991)
2. Preparation and properties of nitrogen oxide: Shak. 2 163

*3 . Reaction of NO with carbon disulfide: Shak. 1 117

*4. Paramagnetism of liquid oxygen: J. Chem. Educ. 57 373 (1980); Shak. 2 147

5. Reaction of xenon with fluorine: J. Chem. Educ. 43 202 (1966)

* See also Shakhashiri Videotapes, Demonstration 7, 38

QUIZZES

Quiz 1

1. For each of the following species, draw the Lewis structure,

describe the geometry, and state whether the species is a dipole.

a. NO_3^- b. SCN^- c. SO_2

2. State the hybridization shown by carbon in

a. CH_4 b. CO_2 c. CH_2O

Quiz 2

1. Draw the Lewis structure, give the hybridization of the central atom, and predict the geometry of:

a. SeF_6 b. XeF_2 c. H_2O

2. Draw the Lewis structure of the C_3H_6 molecule and state all the bond angles.

Quiz 3

1. Draw the Lewis structure and give the hybridization of each atom in:

a. CO b. $COCl_2$ c. F_2O

2. Draw three resonance forms for the CO_3^{2-} ion.

Quiz 4

1. Give the hybridization of sulfur in

a. H_2S b. SO_2 c. SF_6

2. State the ideal bond angles for each species in (1) and indicate whether it is a dipole.

Quiz 5

1. State the number of sigma and pi bonds in

a. CO_3^{2-} b. BF_3 c. CN^- d. NO_2^-

2. Describe the geometry of each species in (1).

Answers

Quiz 1 1. a. :Ö - N - Ö: (with =O: below N) triangular planar; no

b. :S̈ = C = N̈: linear; yes

c. O = S̈ - O: bent; yes

2. a. sp^3 b. sp c. sp^2

Quiz 2.

1. a. SeF_6 (Se bonded to six F) — sp^3d^2; octahedral

b. F - Xe - F — sp^3d; linear

c. H - O - H (bent) — sp^3; bent

2. $H_3C - CH = CH - H$

```
    H   H   H
    |   |   |
H - C - C = C - H
    |
    H
```

120° around double bond, otherwise 109°

Quiz 3 1. a. :C ≡ O: sp

b. Cl - C(=O) - Cl — C is sp^2, O is sp^2, Cl is sp^3

c. F - O - F — sp^3

2. O = C(- O)(- O) ↔ O - C(= O) - O ↔ O - C(- O) = O

Quiz 4 1. a. sp^3 b. sp^2 c. sp^3d^2

2. 109°; yes b. 120°; yes c. 90°, 180°; no

Quiz 5 1. a. 3 sigma, 1 pi b. 3 sigma c. 1 sigma, 2 pi
d. 2 sigma, 1 pi

2. a. triangular planar b. triangular planar
c. linear d. bent

PROBLEMS

1. a. Cl - P(=O...)(- O)(- Cl) - Cl b. F - N(- F) - F c. [O - Cl(- O)(- O) - O]$^-$

d. 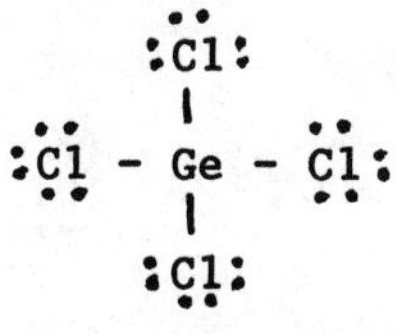

3. a. 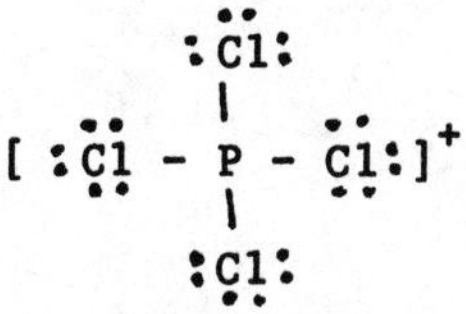 b. 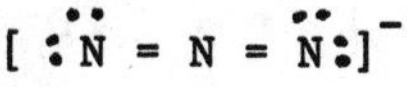 c. 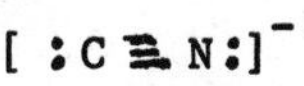

d. 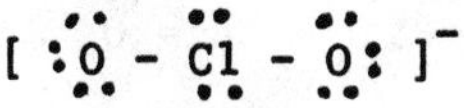

5. a. 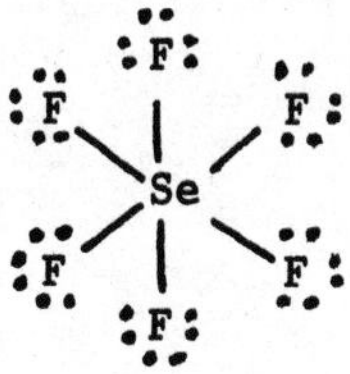 b. 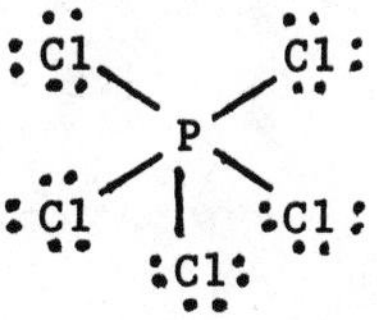 c. 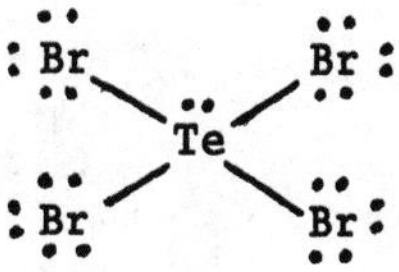

d. 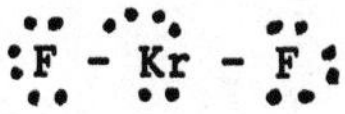

7. a. 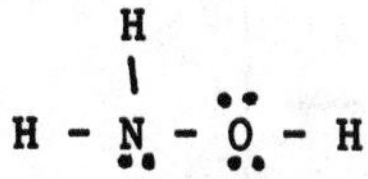 b. 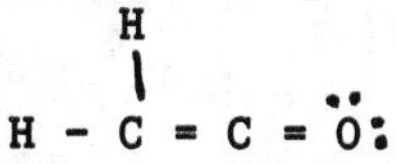 c. 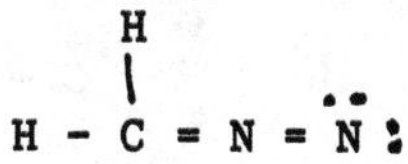

9. 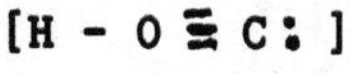

11. 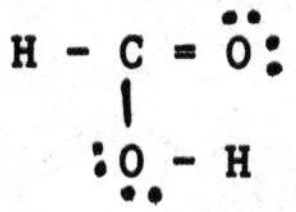

13. 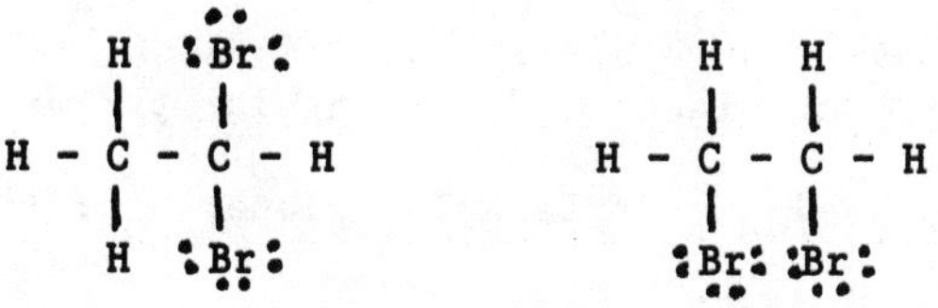

15. a. OH^- b. O_2^{2-} c. CN^- d. SO_4^{2-}

17. a. 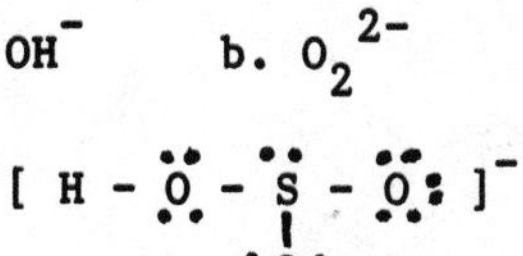 b. 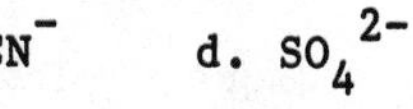 c 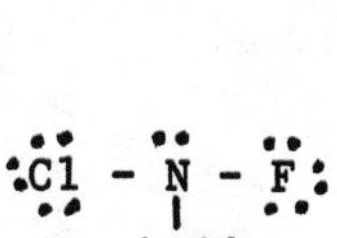

d. 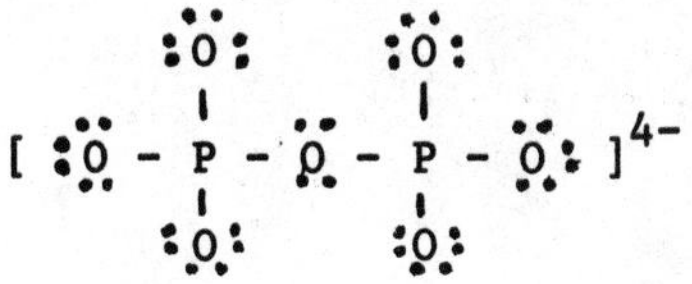

19. a. H - C(H)(H)· b. [:C = O:]$^-$ c. ·N = O: d. :Cl - B(Cl) - Cl:

21. a. [:O - C(=O) - O:]$^{2-}$ ↔ [:O = C(-O) - O:]$^{2-}$ ↔ [:O - C(-O) = O:]$^{2-}$

b. :O - Se(=O) - O: ↔ :O = Se(-O) - O: ↔ :O - Se(-O) = O:

c. [:S - C(=S) - S:]$^{2-}$ [:S = C(-S) - S:]$^{2-}$ [:S - C(-S) = S:]$^{2-}$

23. a. [O, O=C - C(=O), O]$^{2-}$

b. [ ]$^{2-}$ ↔ [ ]$^{2-}$ ↔ [ ]$^{2-}$

c. no; it's an isomer

25. :N(S)(S)N: ↔ :N(S)(S)N: ↔ :N(S)(S)N: ↔ :N(S)(S)N:

27. a. bent b. linear c. linear d. triangular pyramid

29. a. bent b. tetrahedral c. triangular planar d. linear

31. a. RnF_4 (F, F, Rn, F, F) square planar

b. triangular bipyramid c. octahedral

d. $TeBr_4$ (Br, Br, Te, Br, Br) see-saw

33. a. 120° b. 109.5° around C at left; otherwise 180° c. 109.5

35. a. H - O - Cl b. [:O - C ≡ N:]$^-$ c. F - N(-F) - F

bent linear triangular pyramid

d. [:N = N = N:]$^-$

linear

37. a. AX_3E_2; T-shaped b. AX_6; octahedral c. AX_4E; see-saw

39. a. F - Si(F)(F) - F b. [:O - Br(O) - O:]$^-$ c. H - O - Cl - O:

tetrahedral triangular pyramid bent

d. :Cl - As(Cl) - Cl:

triangular pyramid

41. a. NO_2^-; 1, 2, 120° b. NH_4^+ ; 0, 4, 109.5°

c. NO_3^-; 0, 3, 120° d. SCN^-; 0, 2, 180°

43. $H - C_1H_2 - C_2H_2 - C_3(=O) - O_1 - O_2 - N(=O)(O)$

109.5° around C_1, C_2; 120° around C_3; 109.5° around O_1, O_2; 120° around N

45. a. 120° b. 109.5° c. 180°

47. :Cl - Sn - Cl: unshared pair occupies larger volume

:O - S = O: unshared pair occupies larger volume

49. a, b, c, d

51. a, d

53. 1st and 3rd are polar

55. a. sp^2 b. sp   c. sp   d. sp^3

57. a. sp^2 b. sp^3 c. sp^2 d. sp

59. a. 6, sp^3d^2 b. 5, sp^3d c. 6, sp^3d^2 d. 5, sp^3d

61. a. $[:\ddot{F} - \ddot{Cl} - \ddot{F}:]^-$, 5, sp^3d b. $[:\ddot{Cl} - Ge(Cl)_2 - \ddot{Cl}:]^{2-}$, 5, sp^3d

c. $[:\ddot{Cl} - I(Cl)_2 - \ddot{Cl}:]^-$, 6, sp^3d^2

63. C : sp^2 N: sp^2 O : sp^2 and sp^3

65. a. sp^2 b. sp^3 c. sp   d. sp

67. a. $:\ddot{F} - C(=\ddot{O}) - \ddot{F}:$ (sp^2)   b. $:\ddot{F} - \ddot{O} - \ddot{F}:$ (sp^3)   c. $:\ddot{O} - P(Cl)_2 - \ddot{Cl}:$ (sp^3)

69. a. 3 sigma, 1 pi   b. 3 sigma   c. 1 sigma, 2 pi   d. 2 sigma, 2 pi

71. a. BF_4^- b. CO, C_2H_2 c. NO_2^-, HNO_3 d. SO_2

73. a. $d = P\mathcal{M}/RT = 1.250$ g/L

b. $X_{Ar} = 0.0093/0.7901 = 0.0118$

$X_{N_2} = 0.7808/0.7901 = 0.9882$

$\mathcal{M} = 0.9882(28.02 \text{ g/mol}) + 0.0118(39.95 \text{ g/mol})$

$= 28.16$ g/mol

$d = 1.257$ g/L

75. See p. 190

77. $H - \ddot{O} - C \equiv N:$ H = 0, O = 0, C = 0, N = 0 ; more likely

$H - \ddot{O} = C = \ddot{N}:$ H = 0, O = +1, C = 0, N = -1

79. a. more than eight valence electrons around atom

b. molecular structures differing only in electron distribution

c. electron pair assigned solely to one atom

d. odd number of valence electrons (3, 5, 7, - -) around atom

81. b. 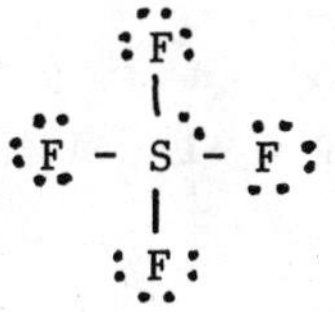 Lewis structure of SF_4: :F: above S; :F: – S – :F: (lone pair on S); :F: below S   d. SF_6

83.

AX_2E_2	2	2	bent	sp^3	polar
AX_3	3	0	triangular planar	sp^2	nonpolar
AX_4E_2	4	2	square planar	sp^3d^2	nonpolar
AX_5	5	0	triangular bipyramid	sp^3d	nonpolar

84. $$n \text{ chlorine fluoride} = \frac{(3.00 \text{ atm})(0.457 \text{ L})}{(0.0821 \text{ L}\cdot\text{atm/mol}\cdot\text{K})(348 \text{ K})} = 0.0480 \text{ mol}$$

$$n\ UF_6 = 5.63 \text{ g } UF_6 \times \frac{1 \text{ mol } UF_6}{352.0 \text{ g } UF_6} = 0.0160 \text{ mol}$$

3 mol chlorine fluoride, 1 mol UF_6

$3ClF_3(g) + U(s) \longrightarrow UF_6(s) + 3ClF(g)$;   x = 3

Lewis structure: :F: above Cl; :F – Cl – F: (two lone pairs on Cl)

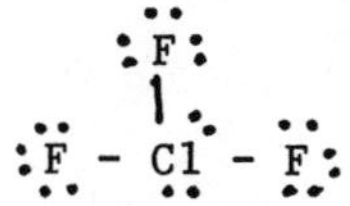

T-shaped; polar; 90°, 180°; sp^3d; 3 sigma

85. 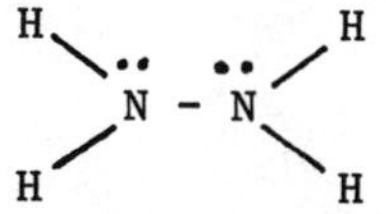 H₂N̈–N̈H₂ (H–N–N–H, each N with one lone pair)   bent, 109.5° bond angle, polar

86. 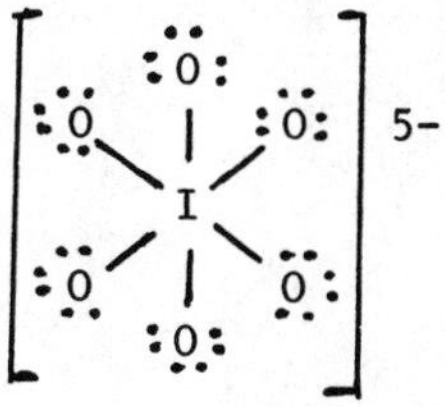 $[IO_6]^{5-}$: I bonded to six :O: atoms   6, sp^3d^2, octahedral

87. a. 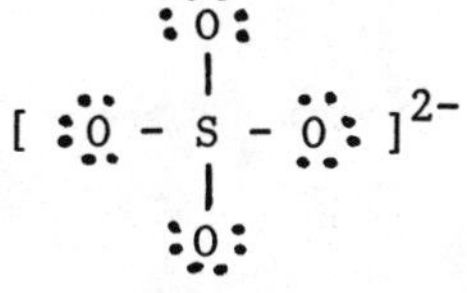 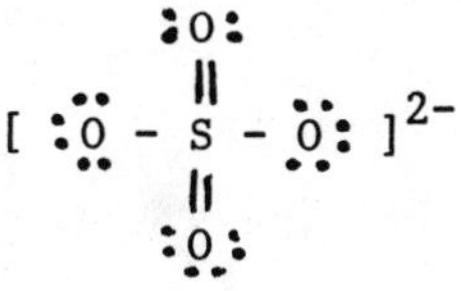 $[\text{:O – S – O:}]^{2-}$ with :O: above and below S (all single bonds); $[\text{:O – S – O:}]^{2-}$ with O= above and below S (double bonds)

b. tetrahedral   c. sp^3

d. 1st structure: S = +2 O = -1

2nd structure: S = 0, O = -1, -1, 0, 0

```
          :O:
           |
88. a. :Cl - P - Cl:       Cl = 0, O = -1, P = +1
           |
          :Cl:

          :O:
           ||
    b. :Cl - P - Cl:
           |
          :Cl:
```

CHAPTER 8
Thermochemistry

LECTURE NOTES

Typically, students find this chapter difficult. In part, this reflects the fact that this is new material, seldom covered in high school. Moreover, thermochemistry is an abstract subject; students have trouble relating it to the real world. For this reason, the experimental aspects of thermochemistry, i.e., calcorimetry, are presented early in the chapter.

In discussing calorimetry (Section 8.2), it is important to point out that

$$q_{reaction} = -q_{calorimeter}$$

i.e., that these two heat flows are equal in magnitude but opposite in sign. Students find the idea of the "calorimeter constant" C hard to grasp. It may help to relate C to the masses and specific heats of the bomb and the water it contains.

Thermochemical equations (Section 8.4) extend the conversion factor approach to heat flow. It is important to emphasize that ΔH is directly proportional to amount of reactant or product. Hess' law is not heavily stressed in this chapter; instead we emphasize the relation

$$\Delta H° = \sum H_f° \text{ products} - \sum H_f° \text{ reactants}$$

Many instructors delete the discussion of the First Law (Section 8.7); it is not required as background in any subsequent chapter. If you cover it, keep in mind that college freshmen seldom appreciate the elegant logic of thermodynamics. Try to relate ΔE and ΔH to real processes.

This chapter requires a minimum of two lectures and could easily expand to three if you spend much time on Section 8.7. In the outlines that follow, we strike an average and include material for 2½ lectures.

LECTURE 1

I **Thermochemistry**

A. Basic concepts

Define and illustrate system, surroundings, state property. Basic equation for heat flow:

$$q = c \times m \times \Delta t \quad (c = \text{specific heat})$$

Suppose 652 J of heat is added to 15.0 g of water (c = 4.18 J/g °C), originally at 20.0°C. Final t?

$$\Delta t = \frac{652\ J}{4.18\ J/g{\cdot}°C \times 15.0\ g} = 10.4°C;\ \text{final } t = 30.4°C$$

Note that if heat is absorbed by system (q positive), temperature increases; if q is negative, temperature drops. For a reaction taking place at constant P and T:

endothermic: $q = \Delta H > 0$ reaction system absorbs heat
exothermic: $q = \Delta H < 0$ reaction system evolves heat

B. Calorimetry

1. Coffee-cup calorimeter ΔH reaction = -q water. That is, heat given off by reaction is absorbed by water in coffee cup.

Suppose heat is absorbed by 412 g of water, increasing its temperature from 20.12 to 29.86°C. What is ΔH?

$$q \text{ water} = 4.18 \frac{J}{g\ °C} \times 412\ g \times 9.74°C = 1.68 \times 10^3\ J$$

$$\Delta H = -1.68\ kJ$$

2. Bomb calorimeter

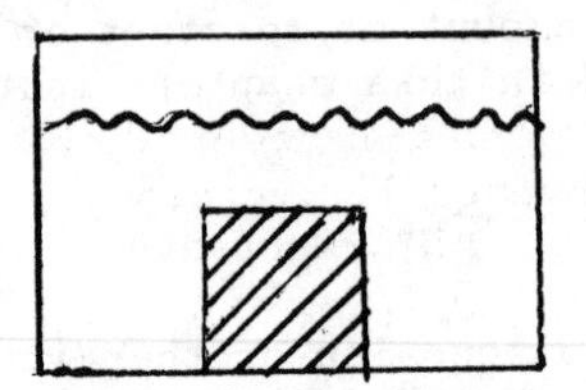

some heat is absorbed by metal as well as water

$$q \text{ reaction} = -q \text{ calorimeter} = -(C \text{ calorimeter}) \times \Delta t$$

where C calorimeter is the total heat capacity of bomb + water

Suppose combustion of 1.60 g CH_4 in bomb calorimter raises temperature by 5.14°C (C = 17.2 kJ/°C)

$$q \text{ reaction} = -17.2\ kJ/°C \times 5.14°C = -88.4\ kJ$$

C. Thermochemical equations specify ΔH in kilojoules

$H_2(g) + Cl_2(g) \longrightarrow 2HCl(g) \quad \Delta H = -185\ kJ$

185 kJ of heat evolved when two moles of HCl are formed

$2HgO(s) \longrightarrow 2Hg(l) + O_2(g) \quad \Delta H = +182\ kJ$

182 kJ of heat must be absorbed to decompose 2 mol HgO

LECTURE 2

I Thermochemistry

A. Rules of thermochemistry

1. ΔH is directly proportional to amount of reactants or products. When one mole of ice melts, 6.00 kJ of heat is absorbed, $\Delta H = +6.00$ kJ. If one gram of ice melts, $\Delta H = 6.00\ kJ/18.02 = +0.333\ kJ$. In general, ΔH can be related to amount by the conversion factor approach.

$$H_2(g) + Cl_2(g) \longrightarrow 2HCl(g) \quad \Delta H = -185\ kJ$$

When 1.00 g of Cl_2 reacts:

$$\Delta H = 1.00\ g\ Cl_2 \times \frac{1\ mol\ Cl_2}{70.90\ g\ Cl_2} \times \frac{-185\ kJ}{1\ mol\ Cl_2} = -2.61\ kJ$$

2. ΔH for a reaction is equal in magnitude but opposite in sign to ΔH for the reverse reaction.

$$H_2O(s) \longrightarrow H_2O(l); \quad \Delta H = +6.00\ kJ; \quad 6.00\ kJ\ absorbed$$

$$H_2O(l) \longrightarrow H_2O(s); \quad \Delta H = -6.00\ kJ; \quad 6.00\ kJ\ evolved$$

3. Hess' law. If Equation 1 + Equation 2 = Equation 3, then

$$\Delta H_3 = \Delta H_1 + \Delta H_2$$

Often used to calculate ΔH for one step, knowing ΔH for all other steps and for the overall reaction.

$$C(s) + \tfrac{1}{2}O_2(g) \longrightarrow CO(g) \quad \Delta H_1 = ?$$

$$CO(g) + \tfrac{1}{2}O_2(g) \longrightarrow CO_2(g) \quad \Delta H_2 = -283.0\ kJ$$

$$C(s) + O_2(g) \longrightarrow CO_2(g) \quad \Delta H_3 = -393.5\ kJ$$

$$\Delta H_1 = -110.5\ kJ$$

B. Heats of formation

1. Meaning. ΔH_f° of compound = ΔH when one mole of compound is formed from the elements in their stable states.

$$2Ag(s) + Cl_2(g) \longrightarrow 2AgCl(s) \quad \Delta H^\circ = -254.0\ kJ$$

$$\Delta H_f^\circ\ AgCl(s) = -127.0\ kJ$$

$HgO(s) \longrightarrow Hg(l) + \frac{1}{2}O_2(g) \quad \Delta H° = +90.8$ kJ

$\Delta H_f°$ $HgO(s)$ = -90.8 kJ

Heats of formation are usually negative; heat is evolved when a compound is formed.

2. Usefulness For any thermochemical equation:

$$\Delta H° = \sum H_f° \text{ products} - \sum H_f° \text{ reactants}$$

Take heat of formation of element in stable state to be zero.

$CH_4(g) + 2O_2(g) \longrightarrow CO_2(g) + 2H_2O(l)$

$\Delta H° = \Delta H_f° \, CO_2(g) + 2\,\Delta H_f° \, H_2O(l) - \Delta H_f° \, CH_4(g)$

$= -393.5$ kJ $+ 2(-285.8$ kJ$) - (-74.8$ kJ$) = -890.3$ kJ

3. Can apply to ions, setting $\Delta H_f°$ $H^+(aq) = 0$

$Zn(s) + 2H^+(aq) \longrightarrow Zn^{2+}(aq) + H_2(g)$

$\Delta H° = \Delta H_f°$ $Zn^{2+}(aq) = -152.4$ kJ

LECTURE 2½

I Bond Energies

B. E. = ΔH when one mole of bonds is broken in gas state

$Cl_2(g) \longrightarrow 2Cl(g) \quad \Delta H$ = B.E. Cl-Cl = 243 kJ

$N_2(g) \longrightarrow 2N(g) \quad \Delta H$ = B.E. N≡N = 941 kJ

In general, multiple bonds are stronger than single bonds

C-C = 347 kJ   C=C = 612 kJ   C≡C = 820 kJ

Estimation of ΔH from bond energies:

$$\Delta H = \sum \text{B.E. bonds broken} - \sum \text{B.E. bonds formed}$$

$N_2(g) + 3H_2(g) \longrightarrow 2NH_3(g)$

ΔH = B.E. N≡N + 3(B.E. H-H) - 6(B.E. N-H) = -85 kJ

(actual ΔH, calculated from heats of formation, is -92 kJ)

II First Law; ΔH and ΔE

A. $\Delta E = q + w$ where ΔE = change in energy of system, q = heat flow into system, w = work done on system.

B. Constant volume, constant pressure processes

constant volume: $w = 0$, $q_V = \Delta E$

constant pressure: $w = -P\Delta V$

$$q_p = \Delta E + P\Delta V = \Delta H$$

DEMONSTRATIONS

1. Endothermic reaction (dissolving ammonium nitrate in water) Test. Dem. 17; J. Chem. Educ. 65 267 (1988); Shak. 1 10
2. Exothermic reactions: Test. Dem. 80
3. Chemical hot pack: Shak. 1 36
4. Cold packs: Shak. 1 8
5. Heats of solution: J. Chem. Educ. 67 426 (1990)

*6. Thermite reaction: Test. Dem. 17, 80, 168; J. Chem. Educ. 42 A 607 (1965), 56 675 (1979); Shak. 1 85

*7. Reaction of hydrogen with oxygen: Test. Dem. 9; J. Chem. Educ. 64 545 (1987); Shak. 1 106

8. Reaction of hydrogen with chlorine: Test. Dem. 10, 154; Shak. 1 121

* See also Shakhashiri Videotapes, Demonstrations 20, 21, 25

QUIZZES

Quiz 1

1. Given the thermochemical equation:

$$SnO_2(s) + 2CO(g) \longrightarrow Sn(s) + 2CO_2(g) \qquad \Delta H = +14.7 \text{ kJ}$$

determine:

a. the amount of heat absorbed when 2.06 g of tin is formed.

b. ΔH for the reaction:

$$2Sn(s) + 4CO_2(g) \longrightarrow 2SnO_2(s) + 4CO(g)$$

c. the heat of formation of SnO_2, given that the heats of formation of CO_2 and CO are -393.5 kJ/mol and -110.5 kJ/mol respectively.

Quiz 2

1. When 2.50 g of potassium chlorate decomposes to potassium chloride and oxygen, 908 J of heat is evolved. Calculate ΔH for the thermochemical equation:

$$2KClO_3(s) \longrightarrow 2KCl(s) + 3\ O_2(g)$$

2. Suppose the 908 J of heat evolved in (1) is absorbed by 45.0 g of water (c = 4.18 J/g·°C). What is the increase in temperature?

Quiz 3

1. Using the table of heats of formation in your text, calculate $\Delta H°$ for the reaction:

$$C_2H_4(g) + 3\ O_2(g) \longrightarrow 2CO_2(g) + 2H_2O(l)$$

2. When 1.00 g of AgCl is formed by the reaction:

$$Ag^+(aq) + Cl^-(aq) \longrightarrow AgCl(s)$$

in a coffee cup calorimeter, the temperature of 20.0 g of water increases from 15.00 to 20.47°C. Taking the specific heat of water to be 4.18 J/g·°C, calculate

a. the amount of heat absorbed by the water.

b. ΔH when one mole of AgCl is formed by this reaction.

Quiz 4

1. Using the table of bond energies in your text, estimate ΔH for the reaction:

$$N_2(g) + 3H_2(g) \longrightarrow 2NH_3(g)$$

2. Given the thermochemical equation:

$$SiO_2(s) + 4HF(g) \longrightarrow SiF_4(s) + 2H_2O(l) \quad \Delta H = -185.8\ kJ$$

and using the table of heats of formation in your text, calculate $\Delta H_f°$ $SiF_4(s)$.

Quiz 5

1. Given the thermochemical equation:

$$CO_3^{2-}(aq) + 2H^+(aq) \longrightarrow CO_2(g) + H_2O(l); \quad \Delta H = -2.2\ kJ$$

calculate:

a. ΔH when 1.40 g of CO_3^{2-} reacts with excess acid.

b. ΔH for the reaction:

$$HCO_3^-(aq) + H^+(aq) \longrightarrow CO_2(g) + H_2O(l)$$

given that

$$CO_3^{2-}(aq) + H^+(aq) \longrightarrow HCO_3^-(aq) \quad \Delta H = -14.9\ kJ$$

c. ΔH for: $CO_2(g) + H_2O(l) \longrightarrow CO_3^{2-}(aq) + 2H^+(aq)$

Answers

Quiz 1 1. a. 0.255 kJ b. -29.4 kJ c. -580.7 kJ/mol

Quiz 2 1. -89.0 kJ 2. 4.83°C

Quiz 3 1. -1410.9 kJ 2. a. 457 J b. -65.5 kJ

Quiz 4 1. -85 kJ 2. -1609.5 kJ/mol

Quiz 5 1. a. -0.051 kJ b. +12.7 kJ c. +2.2 kJ

PROBLEMS

1. a. $C_6H_6(l) \longrightarrow C_6H_6(g) \quad \Delta H = +30.8\ kJ$

   b. $C_{10}H_8(s) \longrightarrow C_{10}H_8(l) \quad \Delta H = +19.3\ kJ$

   c. $Br_2(l) \longrightarrow Br_2(s) \quad \Delta H = -10.8\ kJ$

3. a. $2Al(s) + 3/2\ O_2(g) \longrightarrow Al_2O_3(s)$

   b. $C(s) + 2Cl_2(g) \longrightarrow CCl_4(l)$

   c. $2Na(s) + S(s) + 2\ O_2(g) \longrightarrow Na_2SO_4(s)$

   d. $2K(s) + 2Cr(s) + 7/2\ O_2(g) \longrightarrow K_2Cr_2O_7(s)$

5. $q = 0.902\ \dfrac{J}{g \cdot {}^\circ C} \times 4.75\ g \times 1.8^\circ C = 7.7\ J$

7. $c = 252\ J/(50.0\ g \times 11.4^\circ C) = 0.442\ J/g \cdot {}^\circ C$

9. a. $CaCl_2(s) \longrightarrow Ca^{2+}(aq) + 2Cl^-(aq)$

   b. $q = -4.18\ \dfrac{J}{g \cdot {}^\circ C} \times 100.0\ g \times 4.90^\circ C = -2.05 \times 10^3\ J$

   c. exothermic

   d. $\dfrac{2.05 \times 10^3\ J}{2.80\ g} \times \dfrac{110.98\ g}{1\ mol} = 8.13 \times 10^4\ J = 81.3\ kJ$

11. for 4.50 g: $q = -2.411 \times 10^4\ J/^\circ C \times 3.07^\circ C = -7.40 \times 10^4\ J$

    1 mol : $q = \dfrac{-7.40 \times 10^4\ J}{4.50\ g} \times \dfrac{342.30\ g}{1\ mol} = -5.63 \times 10^6\ J$

13. $q = \dfrac{885.3\ kJ}{16.04\ g} \times 1.750\ g = 96.59\ kJ$

    $C = \dfrac{96.59\ kJ}{3.293^\circ C} = 29.33\ kJ/^\circ C$

15. $q = \frac{5.15 \times 10^3 \text{ kJ}}{1 \text{ mol}} \times \frac{1 \text{ mol}}{128.16 \text{ g}} \times 0.2500 \text{ g} = 10.05 \text{ kJ}$

$\Delta t = \frac{10.05 \times 10^3 \text{ J}}{4.999 \times 10^3 \text{ J/°C}} = 2.01°\text{C}; \; t_f = 22.01°\text{C}$

17. a. 8.16 kJ/1.79°C = 4.56 kJ/°C

b. $q = -4.56 \frac{\text{kJ}}{°\text{C}} \times 10.94°\text{C} = -49.9 \text{ kJ}$

c. $q = \frac{-49.9 \text{ kJ}}{1.00 \text{ g}} \times \frac{26.04 \text{ g}}{1 \text{ mol}} = -1.30 \times 10^3 \text{ kJ}$

19. a. $C_6H_6(l) + 15/2\ O_2(g) \longrightarrow 6CO_2(g) + 3H_2O(l) \quad \Delta H = -3268 \text{ kJ}$

b. exothermic

c. reactants above products

d. $\Delta H = 10.00 \text{ g} \times \frac{3.268 \times 10^3 \text{ kJ}}{78.11 \text{ g}} = -418.4 \text{ kJ}$

e. $1.000 \text{ kJ} \times \frac{78.11 \text{ g}}{3268 \text{ kJ}} = 0.02390 \text{ g}$

21. a. −55.8 kJ

b. $\Delta H = 1.00 \text{ g} \times \frac{-55.8 \text{ kJ}}{18.02 \text{ g}} = -3.10 \text{ kJ}$

23. $\Delta H = 1 \text{ mol } (NH_4)_2Cr_2O_7 \times \frac{252.08 \text{ g}}{1 \text{ mol } (NH_4)_2Cr_2O_7} \times \frac{-1.19 \text{ kJ}}{1.00 \text{ g}}$

$= -3.00 \times 10^2 \text{ kJ}$

a. $(NH_4)_2Cr_2O_7(s) \longrightarrow N_2(g) + Cr_2O_3(s) + 4H_2O(g) \quad \Delta H = -300 \text{ kJ}$

b. $n\ N_2 = \frac{(11.2 \text{ L})(1.00 \text{ atm})}{(0.0821 \text{ L·atm/mol·K})(273 \text{ K})} = 0.500 \text{ mol}$

$\Delta H = -1.50 \times 10^2 \text{ kJ}$

25. a. $C_{57}H_{104}O_6(s) + 80\ O_2(g) \longrightarrow 57CO_2(g) + 52H_2O(l)$

$\Delta H = -3.022 \times 10^4 \text{ kJ}$

b. $q \text{ water} = 1000 \text{ g} \times 5.00°\text{C} \times \frac{4.18 \text{ J}}{\text{g·°C}} = 2.09 \times 10^4 \text{ J}$

$\Delta H = -20.9 \text{ kJ}$

$m = -20.9 \text{ kJ} \times \frac{1 \text{ mol fat}}{-3.022 \times 10^4 \text{ kJ}} \times \frac{885.40 \text{ g}}{1 \text{ mol fat}} = 0.612 \text{ g}$

27. $q_1 = 100.0 \text{ g} \times \frac{1 \text{ mol}}{128.16 \text{ g}} \times \frac{19.3 \text{ kJ}}{1 \text{ mol}} = 15.1 \text{ kJ}$

$q_2 = 100.0 \text{ g} \times \frac{1 \text{ mol}}{18.02 \text{ g}} \times \frac{40.7 \text{ kJ}}{1 \text{ mol}} = 226 \text{ kJ}$

29. a. $q = 25.00 \text{ g} \times 4.18 \text{ J/g·°C} \times 75\text{°C} = 7.8 \times 10^3 \text{ J}$

b. $q = 25.00 \text{ g} \times \frac{1 \text{ mol}}{18.02 \text{ g}} \times \frac{40.7 \text{ kJ}}{1 \text{ mol}} = 56.5 \text{ kJ}$

c. $q = 7.8 \text{ kJ} + 56.5 \text{ kJ} = 64.3 \text{ kJ}$

31.

$PbS(s) + 3/2\ O_2(g) \longrightarrow PbO(s) + SO_2(g) \quad \Delta H = -415.4 \text{ kJ}$

$PbO(s) + C(s) \longrightarrow Pb(s) + CO(g) \quad \Delta H = +108.5 \text{ kJ}$

$PbS(s) + 3/2\ O_2(g) + C(s) \longrightarrow Pb(s) + SO_2(g) + CO(g)$

$\Delta H = -306.9 \text{ kJ}$

33.

$2HNO_3(l) \longrightarrow N_2O_5(g) + H_2O(l) \quad \Delta H = +73.7 \text{ kJ}$

$N_2(g) + 3\ O_2(g) + H_2(g) \longrightarrow 2HNO_3(l) \quad \Delta H = -348.2 \text{ kJ}$

$H_2O(l) \longrightarrow H_2(g) + \frac{1}{2}O_2(g) \quad \Delta H = +285.8 \text{ kJ}$

$N_2(g) + 5/2\ O_2(g) \longrightarrow N_2O_5(g) \quad \Delta H = +11.3 \text{ kJ}$

35. a. $\Delta H_f^\circ = -314.6 \text{ kJ}/2 = -157.3 \text{ kJ}$

b. $\Delta H^\circ = 13.58 \text{ g CuO} \times \frac{1 \text{ mol CuO}}{79.55 \text{ g CuO}} \times \frac{-157.3 \text{ kJ}}{1 \text{ mol CuO}} = -26.85 \text{ kJ}$

37. $CaCO_3(s) \longrightarrow CaO(s) + CO_2(g)$

$\Delta H^\circ = \Delta H_f^\circ\ CaO(s) + \Delta H_f^\circ\ CO_2(g) - \Delta H_f^\circ\ CaCO_3(s) = +178.3 \text{ kJ}$

per gram: $\Delta H^\circ = 1.000 \text{ g } CaCO_3 \times \frac{1 \text{ mol } CaCO_3}{100.09 \text{ g } CaCO_3} \times \frac{178.3 \text{ kJ}}{1 \text{ mol } CaCO_3}$

$= +1.781 \text{ kJ}$

39. a. $\Delta H^\circ = 2\ \Delta H_f^\circ\ I^-(aq) - 2\ \Delta H_f^\circ\ Br^-(aq) = +132.8 \text{ kJ}$

b. $\Delta H^\circ = \Delta H_f^\circ\ CrO_4^{2-}(aq) + 4\ \Delta H_f^\circ\ H_2O(l) + 3\ \Delta H_f^\circ\ Fe^{2+}(aq)$

$- 5\ \Delta H_f^\circ\ OH^-(aq) - \Delta H_f^\circ\ Cr(OH)_3(s) - 3\ \Delta H_f^\circ\ Fe^{3+}(aq)$

$= +67.8 \text{ kJ}$

c. $\Delta H^\circ = \Delta H_f^\circ\ Ni^{2+}(aq) + \Delta H_f^\circ\ SO_2(g) + 2\ \Delta H_f^\circ\ H_2O(l) - \Delta H_f^\circ\ SO_4^{2-}$

$= -13.1 \text{ kJ}$

41. a. $C_2H_5OH(l) + 7/2\ O_2(g) \longrightarrow 2CO_2(g) + 3H_2O(l)$

$\Delta H° = 2\Delta H_f^\circ\ CO_2(g) + 3\Delta H_f^\circ\ H_2O(l) - \Delta H_f^\circ\ C_2H_5OH(l)$
$= -1366.7$ kJ

b. $C_2H_5OH(l) + 7/2\ O_2(g) \longrightarrow 2CO_2(g) + 3H_2O(g)$

$\Delta H° = 2\Delta H_f^\circ\ CO_2(g) + 3\Delta H_f^\circ\ H_2O(g) - \Delta H_f^\circ\ C_2H_5OH(l)$
$= -1234.7$ kJ

43. a. $CaCO_3(s) + 2NH_3(g) \longrightarrow CaCN_2(s) + 3H_2O(l)$; $\Delta H° = +90.1$ kJ

b. $\Delta H_f^\circ\ CaCN_2(s) = 90.1\text{ kJ} - 3\Delta H_f^\circ\ H_2O(l) + \Delta H_f^\circ\ CaCO_3(s) +$
$2\Delta H_f^\circ\ NH_3(g) = -351.6$ kJ

45. $-1936.2\text{ kJ} = 2\Delta H_f^\circ\ Mn^{2+}(aq) + 5\Delta H_f^\circ\ Co^{2+}(aq) + 8\Delta H_f^\circ\ H_2O(l)$
$- 2\Delta H_f^\circ\ MnO_4^-(aq)$

Solving, $\Delta H_f^\circ\ Co^{2+}(aq) = -58.2$ kJ

47. $C_6H_{12}O_6(s) \longrightarrow 2C_2H_5OH(l) + 2CO_2(g)$

$\Delta H° = 2\Delta H_f^\circ\ C_2H_5OH(l) + 2\Delta H_f^\circ\ CO_2(g) + 1275.2\text{ kJ} = -67.2$ kJ

$$\Delta H = 1.000\text{ L} \times 0.120 \times 789\ \frac{g}{L} \times \frac{1\text{ mol}}{46.07\text{ g}} \times \frac{-67.2\text{ kJ}}{2\text{ mol}} = -69.1\text{ kJ}$$

49. a. ΔH = B.E. Cl-Cl = +243 kJ

b. ΔH = 6 B.E. C-H + B.E. C-C = +2831 kJ

c. ΔH = 4 B.E. C-Cl = +1324 kJ

51. $\frac{1}{2}H_2(g) + \frac{1}{2}Br_2(g) \longrightarrow HBr(g)$

$$\Delta H = \frac{\text{B.E. H-H}}{2} + \frac{\text{B.E. Br-Br}}{2} - \text{B.E. H-Br} = -53\text{ kJ}$$

53. C ≡ O + Cl - Cl ⟶ Cl - C - Cl (with C=O: C double-bonded to O below)

ΔH = B.E. C≡O + B.E. Cl-Cl - 2B.E. C-Cl - B.E. C=O = -59 kJ

55. ΔH = B.E. C-O + B.E. H-C - B.E. C-C - B.E. O-H = -46 kJ

57. a. $q = \Delta E - w = -18$ J b. $\Delta E = -35\text{ J} + 128\text{ J} = +93$ J

59. $q_P - q_V = P\Delta V = (-1\text{ mol})RT = -2.48$ kJ

61. a. $C_2H_2(g) + 5/2\ O_2(g) \longrightarrow 2CO_2(g) + H_2O(l)$

$\Delta H = -787.0$ kJ - 285.8 kJ - 226.7 kJ = -1299.5 kJ

b. $\Delta E = \Delta H + 3RT/2 = -1299.5$ kJ + 3.7 kJ = -1295.8 kJ

63. outgo = 165 lb x $\frac{1\ kg}{2.20\ lb}$ x 100 $\frac{kJ}{kg}$ x 1.75 = 1.31×10^4 kJ

input = 2.50×10^3 kcal x $\frac{4.184\ kJ}{1\ kcal}$ = 1.05×10^4 kJ

$\Delta E = 0.26 \times 10^4$ kJ = 2600 kJ; yes, yes

65. (6.0 x 17)kJ + (36 x 17)kJ + (8.0 x 38)kJ = 1020 kJ
about 1 hour

67.

$N_2(g) + 2H_2O(g) \longrightarrow N_2H_4(l) + O_2(g)$	$\Delta H° = +534.2$ kJ
$2H_2(g) + O_2(g) \longrightarrow 2H_2O(g)$	$\Delta H° = -483.6$ kJ
$N_2(g) + 2H_2(g) \longrightarrow N_2H_4(l)$	$\Delta H° = +50.6$ kJ

69. 453.6 g x 32 kJ/1 g = 1.45×10^4 kJ

$\frac{1.45 \times 10^4\ kJ}{1.70 \times 10^3\ kJ/h}$ x 4.0 $\frac{mi}{h}$ = 34 mi

71. n NaOH = 0.500 L x 6.00 mol/L = 3.00 mol

$\Delta H = -133.5$ kJ

1.335×10^5 J = 5.00×10^2 g x 4.18 $\frac{J}{g \cdot °C}$ x Δt

$\Delta t = 63.9°C$; $t_{final} = 83.9°C$

73. $2NH_3(g) + 3/2\ O_2(g) \longrightarrow 3H_2O(g) + N_2(g)$

bond energies: ΔH = 6B.E. N-H + 3/2 B.E. O=O - 6B.E. H-O

- B.E. N≡N = -644 kJ

heats of formation: $\Delta H = -633.2$ kJ

74. Energy gasoline = 1 mi x $\frac{1\ gal}{20.0\ mi}$ x $\frac{4qt}{1\ gal}$ x $\frac{1\ L}{1.057\ qt}$ x $\frac{680\ g}{1\ L}$

x 48 kJ/g = 6.2×10^3 kJ

energy foods = 1 mi x $\frac{1\ h}{13\ mi}$ x $\frac{2.0 \times 10^3\ kJ}{0.30}$ = 5.1×10^2 kJ

6.2×10^3 kJ - 0.51×10^3 kJ = 5.7×10^3 kJ

75. a. $6(38.5\ g) \times 0.902\ \frac{J}{g \cdot °C} \times 20.0°C$

$+ 6(12.0\ oz) \times \frac{453.6\ g}{16\ oz} \times \frac{4.10\ J}{g \cdot °C} \times 20.0°C$

$= 4170\ J + 167{,}400\ J = 1.72 \times 10^5\ J$

b. $1.72 \times 10^2\ kJ \times \frac{1\ mol\ ice}{6.00\ kJ} \times \frac{18.02\ g\ ice}{1\ mol\ ice} = 515\ g$

76. room temperature $\approx$ 25°C

take 100 g water: $q = 100\ g \times 25°C \times 4.18\ J/g \cdot °C = 10{,}400\ J$

mass ice required $= 10.4\ kJ \times \frac{1\ mol}{6.00\ kJ} \times \frac{18.02\ g}{1\ mol} = 31\ g$

$V_{water} = \frac{100\ g}{1.00\ g/cm^3} = 100\ cm^3$

$V_{ice} = \frac{31\ g}{0.90\ g/cm^3} = 34\ cm^3$

$f = 34/134 = 0.25$

77. a. $\Delta H = \Delta H^\circ_f\ Al_2O_3(s) - \Delta H^\circ_f\ Fe_2O_3(s) = -851.5\ kJ$

b. $851{,}500\ J = 0.77\ \frac{J}{g \cdot °C} \times 101.96\ g \times \Delta t + 0.45\ \frac{J}{g \cdot °C}$

$\times 11.70\ g \times \Delta t$

Solving, $\Delta t = 6600°C$

c. yes

CHAPTER 9
Liquids and Solids

LECTURE NOTES

This chapter requires more lecture time than most. Many of the concepts are relatively abstract; most are likely to be new to the student. You should allow at least 2½, more likely 3, lectures for this chapter. Among the more difficult topics are the following:

-trends in melting and boiling points. Students have a great deal of difficulty making comparisons bewteen structurally different types of substances (e.g., NaCl vs Ar, CO_2 vs SiO_2). At this point, they should be able to classify a substance as ionic or molecular, given its formula, but many of them still haven't learned to do this. Students should be expected to recognize some common network covalent substances such as C, Si, SiC, and SiO_2.

- the geometry of different types of unit cells. Models using styrofoam balls are a great help here.

- the idea of vapor pressure. Perhaps surprisingly, students have less trouble with problems involving the Clausius-Clapeyron equation than they do with problems that test their understanding of what vapor pressure really means.

LECTURE 1

I Molecular Substances

A. General properties Nonconductors of electricity, often insoluble in water, low melting, low boiling. Molecules are relatively easy to separate from each other because intermolecular forces are weak.

B. Dispersion forces Result from temporary dipoles formed in adjacent molecules. Strength depends upon how readily electrons are dispersed. Increase with molecular size, molar mass. Explains why boiling point of molecular substances ordinarily increases with molar mass (F_2 < Cl_2 < Br_2 < I_2).

C. Dipole forces Electrical attractive forces between + end of one polar molecule and - end of adjacent molecule. Compare NO (bp = -151°C) to N_2(bp = -196°C), O_2(bp = -183°C).

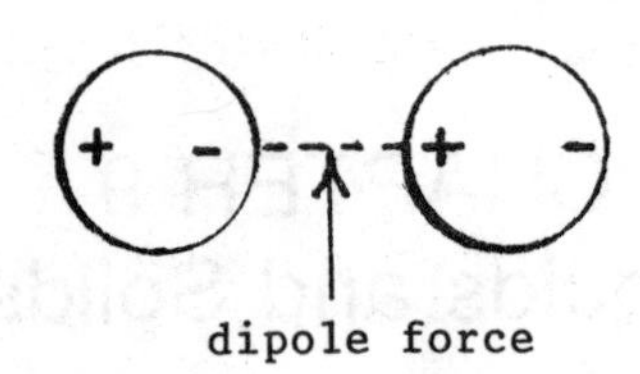

D. Hydrogen bonds Unusually strong dipole force. H atom is very small and differs greatly in electronegativity from F, O, or N. Compare boiling points of Group 16 hydrides:

H_2O (100°C) H_2S(-61°C) H_2Se (-42°C) H_2Te (-2°C)

Note that water has many unusual properties in addition to high boiling point. Open structure of ice, a result of hydrogen bonding, accounts for its low density.

II Other Types of Solids

A. Network covalent (C, SiC, SiO_2)

```
  |   |   |
- X - X - X -
  |   |   |
- X - X - X -
  |   |   |
- X - X - X -
  |   |   |
```

High melting (covalent bonds must be broken). Nonconducting. Insoluble in water or other common solvents. Contrast structure and properties of diamond to those of graphite.

B. Ionic (NaCl, KNO_3)

M^+ X^- M^+

X^- M^+ X^-

M^+ X^- M^+

High melting (strong attractive forces between oppositely charged ions). Nonconducting as solids; conduct molten. Often water soluble. Solubility depends on balance between attractive forces for each other and for water molecules.

C. Metals

M^+ e^- M^+

e^- M^+ e^-

M^+ e^- M^+

"Electron-sea model"; cations in mobile sea of electrons. Conduct electricity. Ductility, malleability, luster. Wide range of melting points, depending upon number of valence electrons. Insoluble in water.

LECTURE 2

I Unit Cells in Metals Unit cell = smallest unit which, repeated over and over again, generates the crystal.

A. Simple cubic Unit cell consists of eight atoms at the corners of a cube.

$$2r = s$$

B. Face-centered cubic Atoms at corners of cube and in center of each face. Atoms touch along face diagonal.

$$4r = s(2)^{1/2}$$

C. Body-centered cubic Atoms at corners of cube and at center of cube. Atoms touch along body diagonal.

$$4r = s(3)^{1/2}$$

Sodium crystallizes in BCC structure; unit cell has length of 0.429 nm. Atomic radius?

$$r = s(3)^{1/2}/4 = 0.186 \text{ nm}$$

II Liquid-Vapor Equilibrium

A. Vapor pressure When a liquid is introduced into a closed container, it establishes equilibrium with its vapor;

$$\text{liquid} \rightleftharpoons \text{vapor}$$

The pressure of the vapor at equilibrium is referred to as the vapor pressure of the liquid.

Vapor pressure is independent of volume of container. Add 0.0100 mol of liquid benzene to 1.00 L flask at 25°C (vp benzene = 92 mm Hg). How much benzene vaporizes?

$$n_{vapor} = \frac{PV}{RT} = \frac{(92/760 \text{ atm})(1.00 \text{ L})}{(0.0821 \text{ L}\cdot\text{atm/mol}\cdot\text{K})(298 \text{ K})} = 0.0050 \text{ mol}$$

$$n_{liquid} = 0.0100 - 0.0050 = 0.0050$$

If the flask were larger than about 2 L, all the liquid would vaporize, equilibrium would not be established.

LECTURE 3

I Liquid-Vapor Equilibrium

A. Temperature dependence of vapor pressure

$$\ln P_2/P_1 = \Delta H_{vap}(1/T_1 - 1/T_2)/R$$

ΔH_{vap} = heat of vaporization in joules per mole

R is the gas law constant in J/mol·K = 8.31 J/mol·K

Take ΔH_{vap} of benzene to be 30.8 kJ/mol, vp = 92 mm Hg at 25°C. Calculate the vapor pressure at 50°C.

$$\ln P_2/P_1 = 30{,}800(1/298 - 1/323)/8.31 = 0.963$$

$P_2/P_1 = 2.62$; $P_2 = 2.62(92 \text{ mm Hg}) = 241 \text{ mm Hg}$

Note that pressure more than doubles when T rises from 25 to 50°C, reflecting the fact that more benzene vaporizes. Pressure of ideal gas would increase by less than 10%.

B. Boiling point = temperature at which vapor bubbles form in liquid.

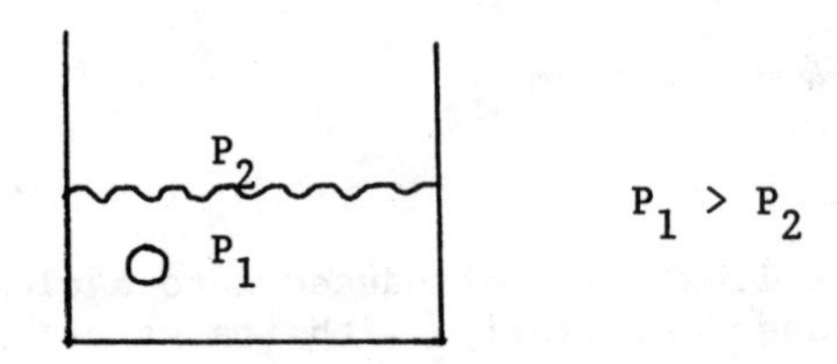

Hence, boiling point varies with applied pressure, P_2. When P_2 = 760 mm Hg, bp water = 100°C. If P_2 = 1075 mm Hg, bp water = 110°C (pressure cooker). If P_2 = 5 mm Hg, water boils at 0°C.

C. Critical temperature Temperature above which liquid cannot exist. Critical pressure = vapor pressure at critical T. Since critical T of oxygen is -119°C, liquid oxygen cannot exist at room temperature, regardless of pressure. Critical T of propane is 97°C; propane is stored as liquid under pressure at room T.

II Phase Diagrams

Graph showing temperatures and pressures at which liquid, solid, and vapor phases of a substance can exist.

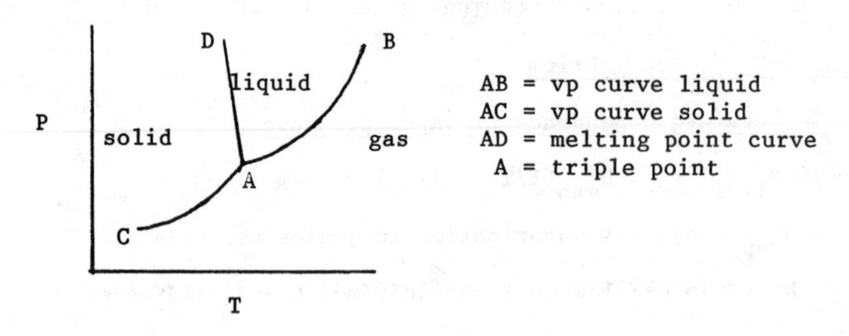

Note that:

- solid sublimes (passes directly to vapor) below triple point (0°C, 5 mm Hg for water; 115°C, 90 mm Hg for iodine)

- if line AD slopes toward P axis, melting point decreases as P increases. This behavior is observed for water, where the liquid is the more dense phase. More often, the solid is more dense, AD tilts away from the P axis, and the melting point increases with pressure.

DEMONSTRATIONS

1. Effect of pressure on boiling point: Shak. 2 81
2. Critical temperature: Test. Dem. 66; J. Chem. Educ. 56 614 (1979), 63 436 (1986), 69 159 (1992)

*3. Liquid carbon dioxide: J. Chem. Educ. 66 597 (1989)

*4. Expansion of water upon freezing: Shak. 3 310

*5. Vapor pressure: Test. Dem. 65; J. Chem. Educ. 58 725 (1981), 64 98 (1987); Shak. 1 6

6. Vapor pressure vs temperature: J. Chem. Educ. 56 474 (1979), 57 667 (1980), 63 629 (1986); Shak. 2 6, 75, 78
7. Phase changes: Test. Dem. 64; J. Chem. Educ. 64 70 (1987)
8. Simultaneous boiling and freezing: J. Chem. Educ. 69 325 (1992)
9. Ice under pressure: J. Chem. Educ. 67 789 (1990)

* See also Shakhashiri Videotapes, Demonstrations 28, 37, 39

QUIZZES

Quiz 1

1. Which has the higher boiling point, NaCl or I_2? CO_2 or SiO_2? Br_2 or I_2? Explain your reasoning.
2. The vapor pressure of a certain liquid doubles when the temperature rises from 12 to 25°C. Calculate its heat of vaporization.

Quiz 2

1. Give two examples of:
   a. network covalent solids
   b. molecules showing hydrogen bonding
   c. molecules with no type of intermolecular force except dispersion.
2. A 0.500 g sample of liquid water is added to a 1.00 L container at 25°C (vp water = 24 mm Hg). Show by calculation whether or not the water completely vaporizes. Explain your reasoning briefly.

Quiz 3

1. A certain metal crystallizes with a body-centered cubic cell 0.318 nm on an edge. What is the atomic radius of the metal?

2. Explain what is meant by
   a. critical temperature   b. hydrogen bond
   c. network covalent solid

Quiz 4

1. Which is stronger
   a. hydrogen bond or covalent bond?
   d. dispersion force in CH4 or in C_2H_6?
   c. ionic bond or dispersion force?
   d. dipole force or polar bond?

2. A certain substance has a triple point at 52°C, 12 mm Hg, and a critical point at 218°C, 18.0 atm. The liquid is less dense than the solid. Draw a phase diagram for the substance and label each area to indicate the phase present.

Quiz 5

1. A certain solid is low melting and almost completely insoluble in water. What structural type of solid might it be (molecular, network covalent, ionic, or metallic)? What further experiments would you carry out to determine the structural type?

2. What is the normal boiling point (1.00 atm) of a liquid which has a vapor pressure of 142 mm Hg at 28°C and a heat of vaporization of 56.2 kJ/mol?

Answers

Quiz 1
1. a. NaCl (ionic)   b. SiO_2 (network covalent)
   c. I_2 (higher molar mass)
2. 38 kJ/mol

Quiz 2
1. a. C, SiC, SiO_2 b. HF, H_2O, NH_3 c. CH_4, Cl_2, H_2
2. $\frac{(0.500/18.02)(0.0821)(298)}{1.00}$ atm = 0.679 atm > vp; no

Quiz 3
1. 0.138 nm
2. a. highest T at which liquid can exist
   b. intermolecular force between H atom and N, O or F atom in adjacent molecule
   c. atoms joined by continuous network of covalent bonds

Quiz 4
1. a. covalent bond   b. C_2H_6 c. ionic bond
   d. polar bond
2. locate and label triple point; liquid-vapor curve ends at 218°C. Liquid-solid line inclines away from P axis.

Quiz 5
1. molecular or metallic; test conductivity
2. 52°C

PROBLEMS

1. He < Ne < Ar < Xe

3. All show dispersion forces; CO shows dipole forces

5. b, c, d

7. a. dipole forces b. higher molar mass c. H bonds in H_2O_2
 d. NaF is ionic

9. none

11. a. CO_2; lower molar mass, nonpolar
 b. HCl; lower molar mass
 c. H_2; lower molar mass
 d. F_2; lower molar mass

13. a. dispersion b. dispersion c. dispersion d. ionic bonds

15. a. probably molecular b. ionic c. metallic

17. a. molecular b. metallic c. molecular

19. a. metallic b. ionic c. molecular d. network covalent
 e. molecular

21. a. $H_2O(s)$, $C_{12}H_{22}O_{11}$, - - b. Na_2O, $CaCO_3$, - - c. SiO_2
 d. $CO_2(s)$

23. a. ions b. molecules c. molecules d. cations, electrons

25. $4(0.143\text{ nm}) = s(2)^{1/2}$; $s = 0.405\text{ nm}$

27. $s = (0.02376\text{ nm})^{1/3} = 0.2875\text{ nm}$

 $4r = (0.2875\text{ nm})(3)^{1/2}$; $r = 0.1245\text{ nm}$

29. a. $2\ r\ K^+ + 2\ r\ I^- = 0.698\text{ nm}$

 b. $0.698\text{ nm} \times (2)^{1/2} = 0.987\text{ nm}$

31. a. $2(0.169\text{ nm}) + 2(0.181\text{ nm}) = 0.700\text{ nm}$

 b. $0.700\text{ nm} = s(3)^{1/2}$; $s = 0.404\text{ nm}$

33. 1;1

35. a. $325 \text{ mm Hg} \times \frac{323 \text{ K}}{353 \text{ K}} = 297 \text{ mm Hg}$

$325 \text{ mm Hg} \times \frac{333 \text{ K}}{353 \text{ K}} = 307 \text{ mm Hg}$

b. 297 mm Hg > 269 mm Hg
307 mm Hg < 389 mm Hg

c. 269 mm Hg at 50°C, 307 mm Hg at 60°C

37. a. $m = \frac{(537/760 \text{ atm})(0.250 \text{ L})(74.12 \text{ g/mol})}{(0.0821 \text{ L-atm/mol}\cdot\text{K})(298 \text{ K})} = 0.535 \text{ g}$

$m_{orig.} = 7.08$ g; yes

b. $V = \frac{(7.08/74.12 \text{ mol})(0.0821 \text{ L}\cdot\text{atm/mol}\cdot\text{K})(298 \text{ K})}{537/760 \text{ atm}} = 3.31 \text{ L}$

c. $P = \frac{3.31 \text{ L}}{5.00 \text{ L}} \times 537 \text{ mm Hg} = 355 \text{ mm Hg}$

39. a. $\ln \frac{90.00}{9.708} = \frac{\Delta H}{8.31}\left[\frac{1}{253.2} - \frac{1}{293.2}\right]$

$\Delta H = 3.43 \times 10^4 \text{ J} = 34.3 \text{ kJ}$

b. $\ln \frac{380}{90.00} = \frac{34300}{8.31}\left[\frac{1}{293.0} - \frac{1}{T}\right]$

$T = 326 \text{ K} = 53°\text{C}$

41. $\ln \frac{760}{589} = \frac{40{,}700}{8.31}\left[\frac{1}{T} - \frac{1}{373.2}\right]$

$T = 366 \text{ K} = 93°\text{C}$

43. $\ln \frac{760}{P} = \frac{59{,}400}{8.31}\left[\frac{1}{373} - \frac{1}{630}\right]$ $P = 0.30$ mm Hg at 100°C

$P = 2.8 \times 10^{-4}$ mm Hg at 0°C

45.

ln P	3.69	4.09	4.38	4.61
1/T	3.28×10^{-3}	3.19×10^{-3}	3.12×10^{-3}	3.08×10^{-3}

slope $\approx -0.92/0.00020 = -4600 = -\Delta H/8.31$

$\Delta H = 3.8 \times 10^4 \text{ J} = 38 \text{ kJ}$

47. a. T b. T c. T d. F

49. a. solid b. liquid c. vapor

51. a. solid b. liquid c. liquid

53.

a.

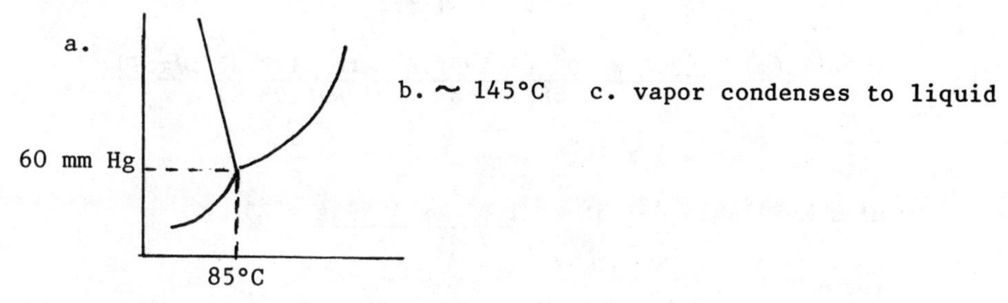

b. ~ 145°C   c. vapor condenses to liquid

55. asbestos can cause lung cancer

57. two-dimensional layer structures

59. a. independent of volume
    b. ln P is a linear function of 1/T
    c. for nonpolar molecular substances
    d. depends on strength of intermolecular forces

61. a. In FCC, there is an atom at the center of each face; in BCC, an atom is at the center of the cube.
    b. solid goes to liquid vs vapor
    c. normal bp is at 1 atm
    d. vapor pressure curve is one part of phase diagram

63. a. $200.6 \text{ g Hg} \times \frac{1 \text{ cm}^3}{13.6 \text{ g Hg}} = 14.8 \text{ cm}^3$

    b. $V = \frac{(1 \text{ mol})(0.0821 \text{ L·atm/mol·K})(293 \text{ K})}{(0.0012/760 \text{ atm})} = 1.5 \times 10^7 \text{ L}$

    c. $V = \frac{4\pi}{3}(1.55 \times 10^{-8} \text{ cm})^3 \times 6.022 \times 10^{23} = 9.39 \text{ cm}^3$

    d. liquid: $\frac{9.39}{14.8} \times 100\% = 63.4\%$

    vapor: $9.39/(1.5 \times 10^{10}) \times 100\% = 6.3 \times 10^{-8}\%$

65. network covalent

67. ethane, carbon disulfide

68. a. 3.60 g water formed

    $m\ H_2O(g) = \frac{(26.74/760 \text{ atm})(10.0 \text{ L})(18.02 \text{ g/mol})}{(0.0821 \text{ L·atm/mol·K})(300 \text{ K})} = 0.257 \text{ g}$

    liquid and vapor

    b. 26.7 mm Hg

69. $V = 12 \text{ ft} \times 13 \text{ ft} \times 8.0 \text{ ft} \times \frac{28.32 \text{ L}}{1 \text{ ft}^3} = 3.5 \times 10^4 \text{ L}$

$$m\ CHCl_3(g) = \frac{(3.5 \times 10^4 \text{ L})(199/760 \text{ atm})(119.37 \text{ g/mol})}{(0.0821 \text{ L·atm/mol·K})(298 \text{ K})}$$

$$= 4.5 \times 10^4 \text{ g}$$

$$\text{mass available} = \frac{1}{4} \text{ qt} \times \frac{1 \text{ L}}{1.057 \text{ qt}} \times \frac{1489 \text{ g}}{1 \text{ L}} = 352 \text{ g}$$

yes

70. $\frac{120 \text{ lb}}{0.10 \text{ in}^3} \times \frac{1 \text{ atm}}{15 \text{ lb/in}^2} = 80 \text{ atm}$

80/134 = 0.60°C; heat conduction is more likely

71. $4\ r\ \text{anion} = (2)^{\frac{1}{2}}(2\ r\ \text{anion} + 2\ r\ \text{cation})$

$$\frac{r\ \text{cation}}{r\ \text{anion}} = \frac{4 - 2(2)^{\frac{1}{2}}}{2(2)^{\frac{1}{2}}} = 0.4144$$

72. $P\ C_3H_8$ is the vapor pressure of propane, which decreases exponentially with T; $P\ N_2$ is gas pressure which decreases linearly with T.

CHAPTER 10
Solutions

LECTURE NOTES

This chapter can be covered in two lectures. The first deals largely with concentration units, the second with colligative properties. Note that molarity was introduced in Chapter 4. Discussed here for the first time are the calculations involved in preparing a solution of known molarity from a more concentrated solution. The important point to get across is that the number of moles of solute stays the same.

Perhaps the most difficult topic in this chapter is the conversion from one concentration unit to another. Students typically don't know where to start. The discussion on pp 267-268 may be helpful here.

LECTURE 1

I Types of Solutes

A. Nonelectrolytes: dissolve as molecules

$CH_3OH(l) \longrightarrow CH_3OH(aq)$

B. Electrolytes: dissolve as ions

$NaCl(s) \longrightarrow Na^+(aq) + Cl^-(aq)$; 2 mol ions per mole solute

$CaCl_2(s) \longrightarrow Ca^{2+}(aq) + 2Cl^-(aq)$ 3 mol ions per mole solute

II Concentrations of Solutes

A. $\text{Mass \% solute} = \frac{\text{mass solute}}{\text{total mass solution}} \times 100$

ppm (liquid solution = Mass % x 10^4

ppb (liquid solution) = mass % x 10^7

B. Mole fraction: $X_a = \frac{\text{no. moles A}}{\text{total number moles}}$

Dissolve 12.0 g CH_3OH in 100.0 g water; mole fraction CH_3OH?

$$n\ CH_3OH = 12.0\ g \times \frac{1\ mol}{32.0\ g} = 0.375$$

$$n\ H_2O = 100.0\ g \times \frac{1\ mol}{18.02\ g} = 5.55\ mol$$

$$X\ CH_3OH = \frac{0.375}{0.375 + 5.55} = 0.0631;\ X\ H_2O = 1 - 0.0631 = 0.9369$$

B. Molality $\quad m = \dfrac{\text{no. moles solute}}{\text{no. kg. solvent}}$

Molality of solution of CH_3OH referred to above?

$m = 0.375/0.1000 = 3.75$ mol/kg solvent

C. Molarity $\quad M = \dfrac{\text{no. moles solute}}{\text{no. liters solution}}$

1. How prepare 35.0 mL of 0.200 M $Al(NO_3)_3$ from solid?

   molar mass = (26.98 + 42.03 + 144.00)g/mol = 213.01 g/mol

   no. moles needed = 0.0350 L x 0.200 mol/L
   = 7.00×10^{-3} mol

   $$\text{mass needed} = 7.00 \times 10^{-3}\ mol \times \frac{213.01\ g}{1\ mol} = 1.49\ g$$

   Dissolve 1.49 g $Al(NO_3)_3$ in enough water to form 35.0 mL of solution.

2. How prepare 35.0 mL of 0.200 M $Al(NO_3)_3$ from 0.500 M solution? Note that number of moles of solute remains constant.

   $$0.0350\ L \times 0.200\ \frac{mol}{L} = 0.500\ \frac{mol}{L} \times V$$

   V = 0.0140 L = 14.0 mL; dilute to 35.0 mL with water

3. Molality of 0.200 M $Al(NO_3)_3$ solution (d = 1.012 g/mL)?

   Work with one liter of solution.

   $$\text{Mass} = 1000\ mL \times \frac{1.012\ g}{1\ mL} = 1012\ g$$

   $$\text{Mass}\ Al(NO_3)_3 = 0.200\ mol \times \frac{213.01\ g}{1\ mol} = 42.6\ g$$

   Mass water = 1012 g - 43 g = 969 g

$$\text{Molality} = \frac{0.200 \text{ mol}}{0.969 \text{ kg}} = 0.206 \text{ mol/kg}$$

LECTURE 2

I Principles of Solubility

A. Nature of solute and solvent Most nonelectrolytes that are appreciably soluble in water are hydrogen bonded (CH_3OH, H_2O_2, sugars). Other types of nonelectrolytes are generally more soluble in nonpolar or slightly polar solvents such as benzene.

B. Effect of temperature
Increase in T favors endothermic process:

solid + water → solution ΔH usually positive, so solubility increases with T

gas + water → solution ΔH usually negative, so solubility decreases with T

C. Effect of pressure
Negligible, except for gases, where solubility is directly proportional to the partial pressure of gas. Carbonated beverages.

II Colligative Properties of Nonelectrolytes
Depend primarily upon concentration of solute particles rather than type.

A. Vapor pressure lowering

$VPL = X_2P_1^o$ where X_2 = mole fraction solute, P_1^o = vapor pressure pure solvent.

B. Osmosis, osmotic pressure

Water moves through semi-permeable membrane from region of high vapor pressure (pure water) to region of low vapor pressure (solution).

$\pi = MRT$; 1 M solution at 25°C has osmotic pressure of 24.5 atm.

C. Boiling point elevation, freezing point lowering
1. Results from vapor pressure lowering.

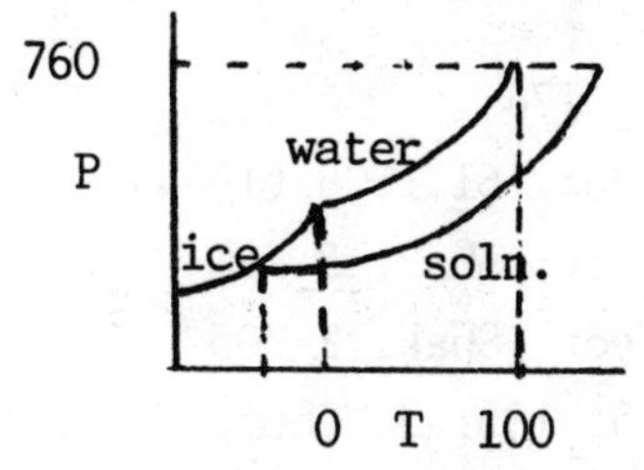

2. For water solutions:

$\Delta T_f = 1.86°C \times m \qquad \Delta T_b = 0.52°C \times m$

Freezing point and boiling point of solution containing 20.0 g of ethylene glycol (molar mass = 62.0 g/mol) in 50.0 g water?

$$m = \frac{20.0/62.0}{0.050} = 6.45$$

$\Delta T_f = 6.45 \times 1.86°C = 12.0°C; \quad T_f = -12.0°C$

$\Delta T_b = 6.45 \times 0.52°C = 3.4°C; \quad T_b = 103.4°C$

3. Use in determining molar mass. Suppose solution is prepared by dissolving 0.100 g of nonelectrolyte in 1.00 g of water; freezing point found to be -1.00°C. Molar mass?

$$\text{molality} = 1.00/1.86 = \frac{0.100/\mathcal{M}}{0.00100}$$

$\mathcal{M} = 0.186/0.00100 = 186$ g/mol

III Electrolytes

Colligative effects are greater because of increased number of particles.

$$\Delta T_f = 1.86°C \times m \times i$$

where i is approximately equal to the number of moles of ions per mole of solute (2 for NaCl, 3 for $CaCl_2$, etc.) Actually, i is usually less than number of moles because of ion atmosphere effects.

DEMONSTRATIONS

*1. Supersaturation: J. Chem. Educ. 57 152 (1980); Shak. 1 27
2. Conductivity of water solutions: Test. Dem. 15, 74, 129; Shak. 3 140
3. Conductivity of HCl in water and benzene: J. Chem. Educ. 43 A539 (1966)
4. Solubility of ammonia: Test. Dem. 174
5. Solubility of iodine: J. Chem. Educ. 61 1009 (1984)
6. Effect of T and P on gas solubility: Shak. 3 280
7. Gas solubility; the fountain effect: Shak. 2 205
8. Raoult's law: Test. Dem. 128, 145, 195; J. Chem. Educ. 53 303 (1976); Shak. 3 242, 254

9. Osmosis: Test. Dem. 16, 66; Shak. 3 283

10. Osmotic pressure of sugar solution: Shak. 3 286

11. Freezing point lowering: Shak. 3 290

12. Freezing point lowering in soda bottle: J. Chem. Educ. 68 1038 (1991)

13. Boiling point elevation: Shak. 3 297

* See also Shakhashiri Videotapes, Demonstration 41

QUIZZES

Quiz 1

1. A solution is prepared by dissolving 5.82 g of urea, $CO(NH_2)_2$, in 24.0 g of water. Calculate the
   a. molality of urea
   b. mole fraction of urea
   c. freezing point of the solution (k_f = 1.86°C/m)
   d. molarity of urea, taking the density of the solution to be 1.010 g/ml.

Quiz 2

1. How would you prepare 250.0 mL of 0.2410 M NaOH from 1.000 M NaOH?
2. A solution containing 1.80 g of a nonelectrolyte in 10.0 g of water freezes at -3.45°C. What is the molar mass of the nonelectrolyte (k_f = 1.86°C/m)

Quiz 3

1. In dilute nitric acid, the concentration of HNO_3 is 6.00 M and the density is 1.19 g/mL. What is:

   a. the mass percent of HNO_3?

   b. the mole fraction of HNO_3?

2. What is the freezing point of a solution containing 12.0 g of naphthalene, $C_{10}H_8$, in 100.0 g of benzene? (k_f = 5.10°C/m, fp pure benzene = 5.50°C)

Quiz 4

1. How would you prepare 218 mL of 0.169 M K_2CrO_4 starting with

   a. pure K_2CrO_4?

   b. a 0.414 M solution of K_2CrO_4?

2. A solution containing 0.125 g of a nonelectrolyte in 10.0 mL of solution has an osmotic pressure of 3.12 atm at 20°C. What is the molar mass of the nonelectrolyte?

Quiz 5

1. A solution is prepared by dissolving 16.0 g of ethylene glycol, $C_2H_6O_2$, in 50.0 g of water. Calculate

   a. the molality of ethylene glycol
   b. the mole fraction of ethylene glycol
   c. the freezing point of the solution (k_f = 1.86°C/m)
   d. the molarity of ethylene glycol (d solution = 0.965 g/mL)

Answers

Quiz 1 1. a. 4.04 m b. 0.0678 c. -7.51°C d. 3.28 M

Quiz 2 1. Mix 60.25 mL of 1.000 M NaOH with enough water to make 250.0 mL of solution.
2. 97.3 g/mol

Quiz 3 1. a. 31.8 b. 0.118 2. 0.72°C

Quiz 4 1. a. Add 7.16 g K_2CrO_4 to enough water to form 218 mL of solution
b. Dilute 89.0 mL of 0.414 M K_2CrO_4 to 218 mL with water.
2. 96.4 g/mol

Quiz 5 1. a. 5.16 m b. 0.0851 c. -9.60°C d. 1.77 M

PROBLEMS

1. a. $C_{12}H_{22}O_{11}$ b. HNO_3 c. $CaCl_2$

3. a. $Cr(NO_3)_3(s) \longrightarrow Cr^{3+}(aq) + 3NO_3^-(aq)$

   b. $Ca_3(PO_4)_2(s) \longrightarrow 3Ca^{2+}(aq) + 2PO_4^{3-}(aq)$

   c. $MgI_2(s) \longrightarrow Mg^{2+}(aq) + 2I^-(aq)$

   d. $RbHCO_3(s) \longrightarrow Rb^+(aq) + HCO_3^-(aq)$

5. a. $C_{12}H_{22}O_{11}(s) \longrightarrow C_{12}H_{22}O_{11}(aq)$

   b. $HCl(g) \longrightarrow H^+(aq) + Cl^-(aq)$

   c. $Al(NO_3)_3(s) \longrightarrow Al^{3+}(aq) + 3NO_3^-(aq)$

   d. $I_2(s) \longrightarrow I_2(aq)$

7. a. $\frac{4.87}{29.87} \times 100\% = 16.3\%$

   b. 83.7%

c. $n\ Na_2Cr_2O_7 = 4.87\ g \times \frac{1\ mol}{261.98\ g} = 0.0186\ mol$

$n\ H_2O = 25.0\ g \times \frac{1\ mol}{18.02\ g} = 1.39\ mol$

$X\ Na_2Cr_2O_7 = \frac{0.0186}{0.0186 + 1.39} = 0.0132$

9. Take 100 mL solution; 45% alcohol by volume

$n\ C_2H_5OH = 45\ mL \times 0.789\ \frac{g}{mL} \times \frac{1\ mol}{46.07\ g} = 0.77$

mass H_2O = 55 mL x 1.00 g/mL = 55 g

molality $= \frac{0.77}{0.055}\ m = 14\ m$

11. mass $Br^- = 58\ mol \times \frac{79.90\ g}{1\ mol} = 4.6 \times 10^3\ g$

$ppm = \frac{4.6 \times 10^3}{1.0 \times 10^6} \times 10^6 = 4.6 \times 10^3$

13. $\mathcal{M}$ = [22.99 + 54.94 + 64.00]g/mol = 141.93 g/mol

	m	n	V	Molarity
a.	10.3 g	0.0726 mol	315 mL	0.230 M
b.	376 g	2.65 mol	3.19 L	0.832 M
c.	863 g	6.08 mol	3.85 L	1.58 M

15. $\mathcal{M}$ = [4.032 + 24.02 + 32.00]g/mol = 60.05 g/mol

	molality	%	ppm	X
a.	0.257 m	1.52%	1.52×10^4	0.00461
b.	0.877 m	5.00%	5.00×10^4	0.0156
c.	0.02572 m	0.1542%	1542	4.632×10^{-4}
d.	35.2 m	67.8%	6.78×10^5	0.387

17. a. mass $= 0.350\ L \times 0.250\ \frac{mol}{L} \times \frac{194.2\ g}{1\ mol} = 17.0\ g$

Dissolve 17.0 g K_2CrO_4 to form 350.0 mL of solution

b. $0.350\ L \times 0.250\ \frac{mol}{L} = V \times 1.50\ \frac{mol}{L}$; V = 0.0583 L

Dilute 58.3 mL of 1.50 M solution to 350 mL

19. a. $0.175\ L \times 0.238\ \frac{mol}{L} = 0.500\ L \times [Al(NO_3)_3]$

$[Al(NO_3)_3] = 0.0833$ M: $[Al^{3+}] = 0.0833$ M; $[NO_3^-] = 0.250$ M

b. $0.175 \text{ L} \times \frac{0.238 \text{ mol } Al(NO_3)_3}{1 \text{ L}} \times \frac{3 \text{ mol } NO_3^-}{1 \text{ mol } Al(NO_3)_3} = 0.125 \text{ mol}$

21. a. Consider 100 g of solution

$n\ H_2SO_4 = 98.0 \text{ g} \times \frac{1 \text{ mol}}{98.09 \text{ g}} = 0.999 \text{ mol}$

$V = 100 \text{ g} \times \frac{1 \text{ mL}}{1.83 \text{ g}} = 54.6 \text{ mL}$

$M\ H_2SO_4 = 0.999 \text{ mol}/0.0546 \text{ L} = 18.3$ M

b. $1.50 \text{ L} \times 3.00 \frac{\text{mol}}{\text{L}} = 18.3 \frac{\text{mol}}{\text{L}} \times V$

$V = 0.246$ L; dilute 246 mL conc. H_2SO_4 to 1.50 L

23. a. One liter solution

n KOH = 1.13 mol; mass KOH = 63.4 g

mass H_2O = 1050 g - 63.4 g = 987 g

$\text{molality} = \frac{1.13 \text{ mol}}{0.987 \text{ kg}} = 1.14 \text{ m}$

$\text{mass } \% = \frac{63.4 \text{ g}}{1050 \text{ g}} \times 100\% = 6.04\%$

b. 30.0 g KOH, 70.0 g water

$n \text{ KOH} = 30.0 \text{ g} \times \frac{1 \text{ mol}}{56.11 \text{ g}} = 0.535 \text{ mol}$

$V = 100.0 \text{ g} \times \frac{1 \text{ mL}}{1.29 \text{ g}} = 77.5 \text{ mL}$

Molarity KOH = 0.535 mol/0.0775 L = 6.90 M

molality = 0.535 mol/0.0700 kg = 7.64 m

c. 14.2 mol KOH, 1000 g water

mass KOH = 797 g

$V = 1797 \text{ g} \times \frac{1 \text{ mL}}{1.43 \text{ g}} = 1260 \text{ mL}$

Molarity KOH = 14.2 mol/1.26 L = 11.3 M

$\text{Mass } \% = \frac{797 \text{ g}}{1797 \text{ g}} \times 100\% = 44.4\%$

25. a. CH_3OH; H bonds with H_2O

b. KCl; ionic vs molecular

c. HF; H bonds with H_2O

d. H_2O_2: H bonds with H_2O

27. a. $\Delta H = \Delta H_f^\circ\ NH_4^+ + \Delta H_f^\circ\ Cl^- - \Delta H_f^\circ\ NH_4Cl = +414.7$ kJ
 b. increase

29. 0.12 x 0.0932 M = 0.011 M

31. a. 19.83 mm Hg x 0.0100 = 0.0198 mm Hg

 b. 1.98 mm Hg c. 2.38 mm Hg

 19.63 mm Hg, 17.85 mm Hg, 17.45 mm Hg

33. 0.0273 mol $C_{10}H_8$, 0.3681 mol C_6H_6

 $X\ C_6H_6 = \frac{0.3681}{0.3954} = 0.9310$; vp = 0.9310 x 74.7 mm Hg = 69.5 mm Hg

35. $\mathcal{M}\ CO(NH_2)_2$ = 60.06 g/mol

 a. $[CO(NH_2)_2] = \frac{10.0\ g}{1\ L} \times \frac{1\ mol}{60.06\ g} = 0.167\ mol/L$

 $\pi = 0.167\ \frac{mol}{L} \times 0.0821\ \frac{L\cdot atm}{mol\cdot K} \times 298\ K = 4.09\ atm$

 b. 20.5 atm c. 40.9 atm

37. a. n solute = 0.329 mol

 $\Delta T_f = 1.86\ \frac{°C}{m} \times \frac{0.329}{0.250}\ m = 2.45°C$; $T_f = -2.45°C$

 $\Delta T_b = 0.52\ \frac{°C}{m} \times \frac{0.329}{0.250}\ m = 0.68°C$; $T_b = 100.68°C$

 b. n solute = 0.618 mol

 $\Delta T_f = 3.54°C$; $T_f = -3.54°C$

 $\Delta T_b = 0.99°C$; $T_b = 100.99°C$

39. n solute = 20.0 mL x 0.785 $\frac{g}{mL} \times \frac{1\ mol}{60.09\ g}$ = 0.261 mol

 molality = 0.261/0.0800 = 3.26 m

 $\Delta T_f = 6.06°C$; $T_f = -6.06°C$

 $\Delta T_b = 1.7°C$; $T_b = 101.7°C$

41. n $C_{10}H_8$ = 0.09753 mol; molality = 0.9753 m

	ΔT_f	T_f	ΔT_b	T_b
a.	6.9°C	46.2°C	6.0°C	180.1°C
b.	4.97°C	0.53°C	2.47°C	82.57°C
c.	19.7°C	-13.2°C	2.68°C	83.40°C

43. $\Delta T_f = 6.5°C$; molality $= \dfrac{3.16/\mathcal{M}}{0.0750 \times 0.799} = \dfrac{54.1}{\mathcal{M}}$

$$\frac{54.1}{\mathcal{M}} \times 20.2 = 6.5 \; ; \; \mathcal{M} \approx 170$$

$$n\,C = 0.429 \times \frac{170}{12.01} = 6 \qquad n\,H = 0.024 \times \frac{170}{1.008} = 4$$

$$n\,N = 0.166 \times \frac{170}{14.01} = 2 \qquad n\,O = 0.381 \times \frac{170}{16.00} = 4$$

$$C_6H_4N_2O_4$$

45. $6.4 = 7.1 \times \dfrac{1.52}{\mathcal{M} \times 0.0100}$; $\mathcal{M} = 1.7 \times 10^2$ g/mol

47. [glucose] $= \dfrac{7.7 \text{ atm}}{0.0821 \text{ L·atm/mol·K} \times 298 \text{ K}} = 0.31$ M

49. $\dfrac{3.86}{760}$ atm $= \dfrac{0.270/\mathcal{M}}{0.0500 \text{ L}} \times 0.0821 \dfrac{\text{L-atm}}{\text{mol·K}} \times 298$ K;

$= 2.60 \times 10^4$ g/mol

51. a > d > b > c

53. $\Delta T_f = 0.38°C = 1.86°C \times 0.10 \times i$; $i = 2$; b

55. molarity $\approx \dfrac{50/342.30 \text{ mol}}{1.0 \text{ L}} = 0.15$ M

$$= 0.15 \frac{\text{mol}}{\text{L}} \times 0.0821 \frac{\text{L-atm}}{\text{mol·K}} \times 283 \text{ K} = 3.4 \text{ atm}$$

$$h = 3.4 \text{ atm} \times \frac{76.0 \text{ cm}}{1 \text{ atm}} \times 13.6 = 3500 \text{ cm} = 35 \text{ m}$$

57. n sugar $= 660 \text{ g} \times \dfrac{1 \text{ mol}}{342.30 \text{ g}} = 1.9$ mol

mass water $\approx$ 15 kg

molality $= \dfrac{1.9}{15}$ m $= 0.13$ m

59. Bubble CO_2 through water; heat carefully

61. a. $i = 3$ for $CaCl_2$, 2 for $MgSO_4$
    b. ions free to move in water
    c. ΔH solution is usually positive; heat is absorbed to break down crystal lattice
    d. must exceed osmotic pressure

63. a. must have same osmotic pressure as blood

b. higher concentration of oxygen

c. carbon dioxide bubbles out of solution when pressure drops

d. m depends on kilograms of water, M on liters of solution

65. a. not if solubility is low

b. usually increases

c. can vary widely in concentrated solution

d. i = 3 for $CaCl_2$, 2 for KCl

e. i = 2 for NaCl, 1 for sucrose

67. a. $\text{molarity} = \frac{275.9/342.30 \text{ mol}}{1 \text{ L}} = 0.8060 \text{ M}$

b. mass of water = 1104 g - 276 g = 828 g

$\text{molality} = \frac{275.9/342.30 \text{ mol}}{0.828 \text{ kg}} = 0.973 \text{ m}$

c. $n \text{ sugar} = 275.9 \text{ g} \times \frac{1 \text{ mol}}{342.30 \text{ g}} = 0.8060 \text{ mol}$

$n \text{ water} = 828 \text{ g} \times \frac{1 \text{ mol}}{18.02 \text{ g}} = 45.9 \text{ mol}$

X water = 0.9827; vp = 17.24 mm Hg

69. $\pi = 0.60 \frac{\text{mol}}{\text{L}} \times 1.9 \times 0.0821 \frac{\text{L·atm}}{\text{mol·K}} \times 298 \text{ K} = 28 \text{ atm}$

70. n KOH = 0.2819 mol

$\text{no. kg water} = \frac{0.2819}{0.250} = 1.128$; 1128 g water

mass water available = 113 g - 15.82 g = 97 g

Add 1030 g water

71. Consider one liter of solution containing n moles of solute, density d

$\text{molarity} = n; \quad \text{molality} = \frac{n}{1000d - n\mathcal{M}} = \frac{\text{molarity}}{d - \frac{\mathcal{M} \cdot \text{molarity}}{1000}}$

In dilute solution, d → 1.00 g/mL; molarity → 0, m → M

72. Let x = mass of X

$$\frac{\frac{x}{410} + \frac{0.100 - x}{342}}{0.00100} = \frac{0.500}{1.86}$$

$x = 0.49$ g; 49% X

73. $\frac{142 \text{ g} \times 0.30 \times 0.15 \times 2}{7.0 \times 10^3 \text{ cm}^3} = 1.8 \times 10^{-3} \text{ g/cm}^3$

74. a. Molarity NaOH = $\frac{49.92 \text{ g}}{0.600 \text{ L}} \times \frac{1 \text{ mol}}{40.00 \text{ g}} = 2.08$ M

b. If NaOH is limiting:

$$n\ H_2 = 1.248 \text{ mol NaOH} \times \frac{3 \text{ mol } H_2}{2 \text{ mol } OH^-} = 1.872 \text{ mol } H_2$$

If Al is limiting:

$$n\ H_2 = 41.28 \text{ g Al} \times \frac{1 \text{ mol Al}}{26.98 \text{ g Al}} \times \frac{3 \text{ mol } H_2}{2 \text{ mol Al}} = 2.295 \text{ mol } H_2$$

$$n\ H_2 = 1.872 \text{ mol}$$

c. $V = \frac{(1.872 \text{ mol})(0.0821 \text{ L atm/mol K})(298 \text{ K})}{(734.8/760 \text{ atm})} = 47.4$ L

75. $V = nRT/P$ Henry's Law: n/P = constant

V = constant x RT

CHAPTER 11
Rate of Reaction

LECTURE NOTES

This chapter is difficult for most students. Concepts such as the rate constant, rate expression, and reaction order are abstract; students have trouble relating chemical kinetics to the real world. Perhaps most difficult of all is the relation between reaction mechanism and order (Section 11.6). Here, students must distinguish carefully between

- the equation for one step in a mechanism and the equation for the overall reaction.

- unstable intermediates and major species (reactants, products.

You should expect to spend at least 2½ lectures on this chapter. If your schedule permits, this could profitably be stretched to 3 lectures.

LECTURE 1

I Reaction Rate

A. Meaning   rate = Δ conc. of species/Δt

$$aA + bB \longrightarrow cC + dD$$

$$\text{rate} = \frac{\Delta[C]}{c\Delta t} = \frac{\Delta[D]}{d\Delta t} = \frac{-\Delta[A]}{a\Delta t} = \frac{-\Delta[B]}{b\Delta t}$$

Minus sign used because concentrations of reactants decrease with time. Concentration will be expressed in molarity; time may be in seconds, minutes, years, etc.

B. Dependence on concentration of reactant(s)

Single reactant, A: rate = $k[A]^m$; k = rate constant
m = order

two reactants, A, B; rate = $k[A]^m \times [B]^n$

m + n = overall order

Generally, m and n are positive integers (1, 2, 3). However, they can be 0 or a fraction such as ½.

Determination of m and k from rate-concentration data:

$$CH_3CHO(g) \longrightarrow CH_4(g) + CO(g)$$

rate	2.0 M/s	0.50 M/s	0.080 M/s
$[CH_3CHO]$	1.0 M	0.50 M	0.20 M

$2.0/0.50 = (1.0/0.50)^m$; m = 2

hence, rate = $k[CH_3CHO]^2$

$k = \text{rate}/[CH_3CHO]^2 = \dfrac{2.0\ M/s}{(1.0\ M)^2} = 2.0(M\text{-}s)^{-1}$

C. First order reaction

$\ln [A]_o/[A] = kt$; $[A]_o$ = original concentration of reactant A

$[A]$ = concentration of reactant A at time t

Suppose k = 0.250/s, $[A]_o$ = 1.00 M. What is the concentration of reactant after 10.0 s?

$\ln [A]_o/[A] = 0.250 \times 10.0 = 2.50$; $[A]_o/[A] = e^{2.50} = 12.2$

$[A] = 1.00\ M/12.2 = 0.0819\ M$

How long does it take for the concentration to drop to one half of its original value?

$[A] = [A]_o/2$; $[A]_o/[A] = 2$

$\ln 2 = kt$; $t_{1/2} = \ln 2/k = 0.693/k = 2.77$ s

Note that, for a first order reaction:

- $t_{1/2}$ is independent of original concentration. It takes as long for the concentration to drop from 1.0 M to 0.50 M as it does to drop from 2.0 M to 1.0 M

- $t_{1/2}$ is inversely related to k. If $t_{1/2}$ is small, k is large and vice versa.

LECTURE 2

I Reactions of Other Orders

Zero order; rate = k; $[A] = [A]_o - kt$

plot of [A] vs t is linear

Second order: rate = $k[A]^2$; $1/[A] - 1/[A]_o = kt$

plot of 1/[A] vs t is linear

II Activation Energy

Reaction occurs by collision, but not every collision leads to reaction. Colliding molecules must possess a certain minimum energy to break reactant bonds, bring about reaction.

Diagram:

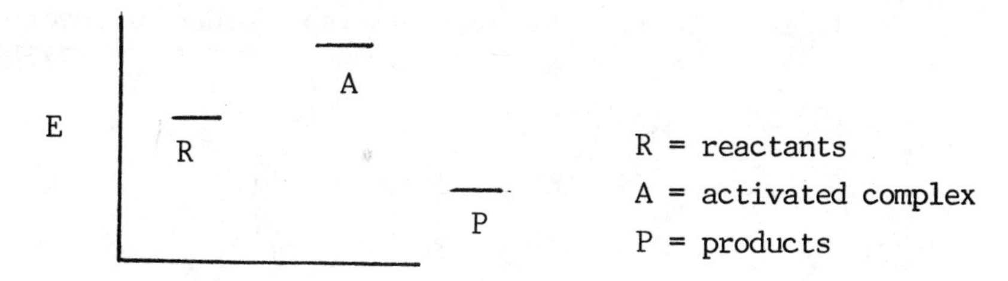

III Effect of Temperature

In general, increase in T increases rate. Rate is approximately doubled when T increases by 10°C. Explanation:

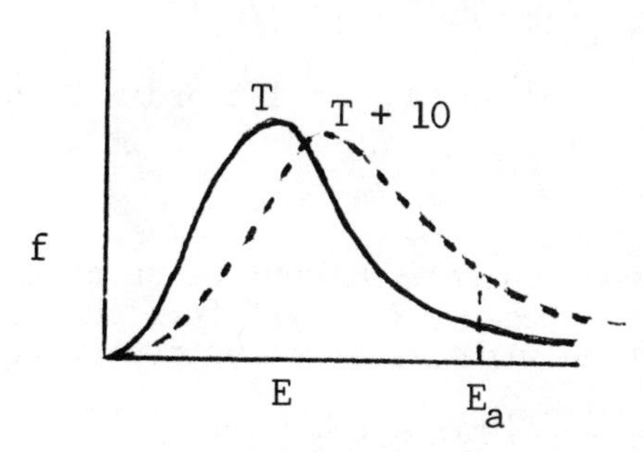

Relation between k and T:

$$\ln k = A - E_a/RT \qquad R = 8.31 \text{ J/K}, E_a \text{ in joules}$$

Plot of ln k vs 1/T is a straight line with slope $-E_a/R$

$$E_a \text{ (in joules)} = -8.31 \times \text{slope}$$

Two point equation:

$$\ln k_2/k_1 = E_a[1/T_1 - 1/T_2]/R$$

Suppose rate doubles when T increases from 25 to 35°C. E_a?

$$\ln k_2/k_1 = 0.693 = E_a[1/298 - 1/308]/8.31$$

$$E_a = 5.3 \times 10^4 \text{ J} = 53 \text{ kJ}$$

LECTURE 2½

I <u>Reaction Mechanism</u>

Most reactions take place in a series of steps. This ordinarily results in a rather complex rate expression. To find the rate expression for a multi-step mechanism:

1. Focus on the slow step; assume rate for that step = overall rate.
2. Coefficients in rate-determining step = order of reaction
3. Eliminate any unstable intermediates from rate expression.

$X_2(g) \rightleftharpoons 2X(g)$ fast

$X(g) + A_2(g) \longrightarrow AX(g) + A(g)$ slow

$A(g) + X_2(g) \longrightarrow AX(g) + X(g)$ fast

rate = $k_2 \times [X] \times [A_2]$

To find [X], note that, for first step:

$$k_1[X_2] = k_1'[X]^2; \quad [X] = (k_1[X_2]/k_1')^{\frac{1}{2}}$$

Hence: rate = $k_2(k_1^{\frac{1}{2}}) \times [X_2]^{\frac{1}{2}} \times [A_2]/k_1'^{\frac{1}{2}}$

$= k \times [X_2]^{\frac{1}{2}} \times [A_2]$; 1st order in A_2, $\frac{1}{2}$ order in X_2

II <u>Catalysis</u>

Catalyst increases reaction rate without being consumed in the reaction. This happens because catalyst furnishes an alternative path for reaction with lower activation energy.

Example: $2H_2O_2(aq) \longrightarrow 2H_2O + O_2(g)$

Direct reaction comes about by collision between H_2O_2 molecules. E_a is high. Reaction is catalyzed by I^- ions:

$H_2O_2 + I^- \longrightarrow H_2O + I^-$

$H_2O_2 + IO^- \longrightarrow H_2O + O_2 + I^-$

$2H_2O_2 \longrightarrow 2H_2O + O_2$

E_a is much lower for this two-step process.

DEMONSTRATIONS

1. Effect of concentration upon rate: Test. Dem. 84; J. Chem. Educ. <u>42</u> A607 (1965); <u>62</u> 153 (1985)

*2. Clock reactions: Test. Dem. 19, 85, 130, 147, 179; J. Chem. Educ. 57 152 (1980), 64 255 (1987); 69 236 (1992); Shak. 4 3-86

*3. Oscillating reactions: Shak. 2 248-307

*4. Catalysis: Test. Dem. 159, 218; J. Chem. Educ. 55 652 (1978), 65 68 (1987)

5. Reaction of hydrogen with chlorine: Shak. 1 121

6. Effect of temperature upon rate: J. Chem. Educ. 58 384 (1981)

7. Kinetics of reduction of MnO_4^-: J. Chem. Educ. 67 598 (1990)

8. Preparation of ozone: Test. Dem. 57, 127

9. Absorption of UV light by ozone: J. Chem. Educ. 66 338 (1989)

* See also Shakhashiri Videotapes, Demonstrations 44, 45, 46, 47, 48

QUIZZES

Quiz 1

1. Given the following data for the reaction between A and B:

[A]	0.100	0.100	0.400
[B]	0.100	0.200	0.200
rate(M/min)	0.080	0.16	0.32

determine:

a. the order of reaction with respect to A
b. the order of reaction with respect to B.
c. the rate constant for the reaction.

Quiz 2

1. For a certain first-order reaction, k = 0.052/h. Calculate
a. the rate when the concentration of reactant is 0.10 M.
b. the half-life
c. the time required for the concentration to drop to 80% of its original value.

Quiz 3

1. For a certain reaction, the activation energy is 62 kJ. Calculate:
a. the ratio of the rate constant at 50°C to that at 0°C.
b. the temperature at which k is 0.025/s, given that k = 0.050/s at 25°C.

Quiz 4

1. Consider the reaction mechanism:

$A + 2C \rightleftharpoons AC_2$ fast

$AC_2 + A \longrightarrow 2AC$ slow

Obtain a rate expression for the reaction of A with C to form AC (the expression should not contain the concentration of AC_2).

2. A certain reaction is zero order in reactant X and second order in reactant Y. If the concentrations of both reactants are doubled, what happens to the reaction rate?

Quiz 5

1. The rate constant, k, for a certain first order reaction is 0.12/min at 20°C.
   a. How long does it take for the concentration of reactant to drop to one fourth of its original value?
   b. What is k at 30°C if the activation energy is 29 kJ?

Answers

Quiz 1 1. a. ½ b. 1 c. 2.5

Quiz 2 1. a. 0.0052 mol/L·h b. 13 h c. 4.3 h

Quiz 3 1. a. 69 b. 17°C

Quiz 4 1. rate = $k[A]^2[C]^2$

2. multiplied by 4

Quiz 5 1. a. 12 min b. 0.18/min

PROBLEMS

1. a. $-\Delta[C_2H_6]/\Delta t$ b. $\Delta[C_2H_4]/\Delta t$

3. 0.40 mol/L·s 0.60 mol/L·s

5. a. rate = $-\Delta[NOCl]/2\Delta t$

   b. rate = $\frac{0.342 \text{ mol/L}}{16.0 \text{ min}}$ = 0.0214 mol/L·min

7. ~0.020 mol/L·min

9. a. 2nd order in A; 1st order in B; 3rd
   b. 1st order in A
   c. 2nd order in A; 2nd order in B; 4th
   d. Zero

11. a. $(mol \cdot L^3)/(L \cdot s \cdot mol^3) = L^2/mol^2 \cdot s$

b. s^{-1}

c. $(mol \cdot L^4)/L \cdot s \cdot mol^4 = L^3/mol^3 \cdot s$

d. mol/L·s

13. rate = 1.5×10^{-3} mol/L·min

$$k = \frac{6.82 \text{ mol/L·min}}{0.035 \text{ mol/L}} = 1.9 \times 10^2/\text{min}$$

$$[A] = \frac{7.98 \times 10^{-3} \text{ mol/L·min}}{0.0372/\text{min}} = 0.215 \text{ mol/L}$$

15. a. rate = $k[NO_2]^2$

b. $0.31 \dfrac{L}{mol \cdot s} \times (0.050 \text{ mol/L})^2 = 7.8 \times 10^{-4}$ mol/L·s

c. $1.5 \times 10^{-3} \dfrac{mol}{L \cdot s} = 0.31 \dfrac{L}{mol \cdot s} [NO_2]^2$; $[NO_2] = 0.070$ mol/L

17. a. $1.6 \times 10^{-8} \dfrac{mol}{L \cdot min} = k(0.020 \text{ mol/L})^2 \times (0.030 \text{ mol/L})$

$k = 1.3 \times 10^{-3}\ L^2/mol^2 \cdot min$

b. $2.0 \times 10^{-6} \dfrac{mol}{L \cdot min} = 1.3 \times 10^{-3} \dfrac{L^3}{mol^2 min} (0.064 \text{ mol/L})^2 \times [Br_2]$

$[Br_2] = 0.38$ M

c. $4.5 \times 10^{-7} \dfrac{mol}{L \cdot min} = 1.3 \times 10^{-3} \dfrac{L^2}{mol^2 min} \times [NO]^2 \times \dfrac{[NO]}{3}$

$[NO]^3 = 3.5 \times 10^{-4}\ mol^3/L^3$; $[NO] = 0.070$ M

19. d

21. rate independent of [A]; zero order

23. a. $\dfrac{2.6 \times 10^{-7}}{1.3 \times 10^{-7}} = (0.20/0.10)^m$ m = 1 1st order in $HgCl_2$

$\dfrac{5.2 \times 10^{-7}}{1.3 \times 10^{-7}} = (0.20/0.10)^n$ n = 2 2nd order in $C_2O_4^{2-}$

3rd order overall

b. rate = $k[HgCl_2] \times [C_2O_4^{2-}]^2$

c. $k = 1.3 \times 10^{-4} L^2/mol^2\ s$ d. rate = 3.5×10^{-6} mol/L s

25. a. $\frac{2.7 \times 10^{-2}}{3.0 \times 10^{-3}} = (0.30/0.10)^m$ m = 2 (2nd order in A)

Zero order in B; 2nd order overall

b. $3.0 \times 10^{-3} \frac{mol}{L\cdot s} = k(0.10\ mol/L)^2$ k = 0.30 L/mol·s

c. $5.0 \times 10^{-3} \frac{mol}{L\cdot s} = 0.30 \frac{L}{mol\cdot s} \times [A]^2$ [A] = 0.13 mol/L

27. a. $\frac{1.26 \times 10^{-2}}{8.21 \times 10^{-3}} = (0.0713/0.0575)^m$ m = 2 2nd order in ClO_2^-

$\frac{1.26 \times 10^{-2}}{8.21 \times 10^{-3}} = (0.0333/0.0216)^n$ n = 1 1st order in OH^-

rate = $k \times [ClO_2^-]^2 \times [OH^-]$

b. $k = \frac{8.21 \times 10^{-3}\ mol/L\cdot s}{(0.0575\ mol/L)^2(0.0216\ mol/L)} = 115\ L^2/mol^2\cdot s$

c. $[OH^-] = \frac{3.56 \times 10^{-2}\ mol/L\cdot s}{(115\ L^2/mol^2\cdot s)(0.100\ mol/L)^2} = 0.0310\ M$

29.

t	0	300	600	900	1200
$[CH_3NO_2]$	0.200	0.145	0.105	0.076	0.055
$\ln[CH_3NO_2]$	-1.609	-1.931	-2.254	-2.577	-2.900
$1/[CH_3NO_2]$	5.00	6.90	9.52	13.2	18.2

Plot of $\ln[CH_3NO_2]$ vs t is linear; 1st order

31. $t_{\frac{1}{2}} = constant/[A]_o$ 2nd order

33. a. 3.2×10^{-6} mol/L·s

b. $k = 4.4 \times 10^{-4}/s$ rate = 3.1×10^{-6} mol/L·s

c. b; $\ln [C_2H_6]$ is linear function of time

35. a. ln 1.00/0.75 = k(80 s); k = 0.0036/s

b. $t_{\frac{1}{2}} = 0.693\ s/0.0036 = 1.9 \times 10^2\ s$

37. a. k = 0.693/64.2 min = 0.0108/min

b. two half-lives: 128.4 min = 2.14 h

c. $\ln \frac{1.000}{0.0500} = \frac{0.0108}{\text{min}} \times t = 3.00$ $\quad t = 278$ min

39. a. [1/0.0104 - 1/0.800] L/mol = k(125 s) k = 0.759 L/mol•s

b. $t_{\frac{1}{2}} = \frac{1}{(0.759 \text{ L/mol}\cdot\text{s})(0.500 \text{ mol/L})} = 2.63$ s

41. [1/0.00200 - 1/0.0200] L/mol = $48 \frac{\text{L}}{\text{mol}\cdot\text{min}} \times t$ $\quad t = 9.4$ min

43. 0.80 mol/L = $0.050 \frac{\text{mol}}{\text{L}\cdot\text{s}} \times t$ $\quad t = 16$ s

45. C fastest, B slowest; rate inversely related to E_a

47. Assigning the reactants zero energy, the products are located at +15 kJ, the activated complex for the uncatalyzed reaction at +120 kJ, and the activated complex for the catayzed reaction at +75 kJ

49.

ln k	-3.04	+0.83	3.89	6.38
1/T	0.00129	0.00115	0.00103	0.00093

slope $\approx -2.6 \times 10^4$ $\quad E_a = 2.2 \times 10^2$ kJ

51. $\ln k_2/k_1$ = 50,200[1/310.2 - 1/313.2]/8.31 = 0.19

$k_2/k_1 = 1.2$; 20%

53. 25°C: X = 148 $\qquad$ 35°C: X = 220

$\ln \frac{220}{148} = E_a[1/298 - 1/308]/8.31 = 0.396$

$E_a = 3.0 \times 10^4$ J = 30 kJ

55. a. ln 0.28/0.0039 = E_a[1/500 - 1/573]/8.31 = 4.3

$E_a = 1.4 \times 10^2$ kJ

b. ln k/0.28 = 140,000[1/573 - 1/673]/8.31 $\quad$ k = 22 L/mol•s

57. $\ln t_{\frac{1}{2}}$/40.8 min = 248,000[1/673 - 1/743]/8.31 = 4.2

$t_{\frac{1}{2}} = 2.7 \times 10^3$ min

59. a. rate = $k[N_2O_2] \times [H_2]$

b. rate = $k[ClCO] \times [Cl_2]$

c. rate = $k[NO_2]^2$

61. $\text{rate} = k_3[H_2] \times [I]^2$

$k_1[I_2] = k_2[I]^2$

$\text{rate} = \frac{k_3[H_2] \times k_1[I_2]}{k_2} \qquad k = k_3k_1/k_2$

63. (1) $\text{rate} = k_3[NO_3] \times [NO]$

$k_1[NO] \times [O_2] = k_2[NO_3]$

$\text{rate} = \frac{k_3k_1}{k_2} \times [NO]^2 \times [O_2]$

(2) $\text{rate} = k_3[N_2O_2] \times [O_2]$

$k_1[NO]^2 = k_2[N_2O_2]$

$\text{rate} = \frac{k_1k_3}{k_2} \times [NO]^2 \times [O_2]$

65. See p. 313

67. Activated complex is 12 kJ lower for the catalyzed reaction. Products are 391 kJ below reactants.

69. $t_{\frac{1}{2}} = \frac{0.693}{3 \times 10^{-26}} \text{ s} \times \frac{1 \text{ min}}{60 \text{ s}} \times \frac{1 \text{ h}}{60 \text{ min}} \times \frac{1 \text{ d}}{24 \text{ h}} \times \frac{1 \text{ yr}}{365.2 \text{ d}}$

$= 7 \times 10^{17}$ yr   not likely

71. a. fewer molecules have activation energy
    b. not necessarily first order
    c. flame supplies activation energy

73. a. $\ln \frac{68.0}{10.0} = \frac{0.037}{\text{min}} \times t \qquad t = 52 \text{ min}$

b. $\ln \frac{68.0}{m} = \frac{0.037}{\text{min}} \times 25 \text{ min} = 0.92 \qquad m = 27 \text{ g}$

41 g of N_2O_5 decompose

$n\ O_2 = 41 \text{ g } N_2O_5 \times \frac{1 \text{ mol } N_2O_5}{108.02 \text{ g } N_2O_5} \times \frac{\frac{1}{2} \text{ mol } O_2}{1 \text{ mol } N_2O_5} = 0.19 \text{ mol}$

$V = \frac{0.19 \text{ mol} \times 0.0821 \text{ L·atm/mol·K} \times 318 \text{ K}}{1.00 \text{ atm}} = 5.0 \text{ L}$

75. $[A]_o - [A] = kt \qquad [A] = [A]_o - kt$

76. $\frac{-d[A]}{adt} = k[A]$; $\frac{-d[A]}{[A]} = akdt$; $\ln[A]_o/[A] = akt$

77. a. $\frac{-d[A]}{dt} = k[A]^2$; $\frac{-d[A]}{[A]^2} = kdt$; $\frac{1}{[A]} - \frac{1}{[A]_o} = kt$

b. $\frac{-d[A]}{[A]^3} = kdt$; $\frac{1}{2[A]^2} - \frac{1}{2[A]_o^2} = kt$

78. Compare 1st and 4th: 2nd order in A

Compare 2nd and 4th: 1st order in C

Compare 1st and 3rd: change in [C] should multiply rate by 2, so must be first order in B

rate = $k[A]^2 \times [B] \times [C]$

79. $a = 0.30 \times \frac{8\ h}{48\ h} = 0.050$; saturation value = $\frac{0.100\ g}{1 - 10^{-0.050}}$

= 0.91 g

$0.500 = \frac{X}{0.11}$ X = 0.055 g

CHAPTER 12
Gaseous Chemical Equilibrium

LECTURE NOTES

You will notice that this chapter deals exclusively with the thermodynamic equilibrium constant (K_p) referred to here simply as K. A major advantage of this approach is that, in future chapters, you don't have to hedge on relations such as that between $\Delta G°$ and K, the temperature dependence of K, etc. It is important to point out (p. 331) that in the expression for K, it is always true that

- gases enter as their partial pressures in atmospheres
- aqueous solutes enter as their molar concentrations
- solids, pure liquids, and the solvent do not appear

Most students find equilibrium calculations difficult to follow, let alone carry out on their own. This is particularly true for the most important calculation: finding equilibrium pressures of all species. Equilibrium tables such as the one shown with Example 12.6 help.

This chapter will require at least 2 lectures, more likely 2½.

LECTURE 1

I The Equilibrium Constant, K

A. $2HI(g) \rightleftharpoons H_2(g) + I_2(g)$

To study this equilibrium at 520°C, put pure HI or a mixture of H_2 and I_2 in a sealed container at this temperature. Take samples over a period of time until the partial pressures become constant. At that point, system is in equilibrium.

	$2HI(g)$	$\rightleftharpoons$ $H_2(g)$	+ $I_2(g)$	
P_o(atm)	0.200	0.000	0.000	
ΔP(atm)	-0.040	+0.020	+0.020	Expt. 1
P_{eq}(atm)	0.160	0.020	0.020	

P_o(atm)	0.100	0.100	0.100	
ΔP(atm)	+0.140	-0.070	-0.070	Expt. 2
P_{eq}(atm)	0.240	0.030	0.030	

Note that:

a. $\Delta P\ HI = -2\Delta P\ H_2 = -2\Delta P\ I_2$

b. $\dfrac{P\ H_2 \times P\ I_2}{(P\ HI)^2} = 0.016$ at equilibrium in all cases

In general, at any T, the quotient, $P\ H_2 \times P\ I_2/(P\ HI)^2$ is a constant, equal to K. The value of K depends only upon temperature; it is independent of original pressures, total pressure, or volume.

II General Expression for K

A. Only gases involved

$$aA(g) + bB(g) \rightleftharpoons cC(g) + dD(g)$$

$$K = \frac{(P\ C)^c \times (P\ D)^d}{(P\ A)^a \times (P\ B)^b}$$

Note that products (right side of equation) appear in numerator, reactants in denominator.

B. Solids or liquids as well as gases

$$CaCO_3(s) \rightleftharpoons CaO(s) + CO_2(g) \qquad K = P\ CO_2$$

Terms for solids or liquids do not appear. Adding or removing a solid or liquid does not affect position of equilibrium. For example, the pressure of CO_2 is the same regardless of how much CaO or $CaCO_3$ is present.

C. Aqueous solutions (covered in later chapters in more detail)

Solutes appear as molarities; solvent H_2O does not appear

$$Ag(s) + 2H^+(aq) + NO_3^-(aq) \longrightarrow Ag^+(aq) + NO_2(g) + H_2O$$

$$K = \frac{[Ag^+] \times P\ NO_2}{[H^+]^2 \times [NO_3^-]}$$

D. Relations between equilibrium constants

$$2HI(g) \rightleftharpoons H_2(g) + I_2(g) \qquad K = \frac{P\ H_2 \times P\ I_2}{(P\ HI)^2} = 0.016$$

$H_2(g) + I_2(g) \rightleftharpoons 2HI(g) \quad K' = \dfrac{(P\ HI)^2}{P\ H_2 \times P\ I_2} = 1/K = 62$

$HI(g) \rightleftharpoons \frac{1}{2}H_2(g) + \frac{1}{2}I_2(g) \quad K'' = \dfrac{(P\ H_2)^{\frac{1}{2}} \times (P\ I_2)^{\frac{1}{2}}}{P\ HI}$

$= K^{\frac{1}{2}} = 0.13$

Rule of multiple equilibria: If Eqn. 3 = Eqn. 1 + Eqn. 2,

$K_3 = K_1 \times K_2$

LECTURE 2

I Calculation of K

$N_2O_4(g) \rightleftharpoons 2NO_2(g)$

At 100°C, $P\ N_2O_4 = 0.50$ atm, $P\ NO_2 = 2.3$ atm at equilibrium

$K = (2.3)^2/0.50 = 11$

II Applications of K

In general, if K is large, equilibrium moves far to right; products favored. If K is small, equilibrium favors reactants.

A. Determination of direction of reaction

Compare the actual pressure quotient, Q, to that required at equilibrium, K. If $Q < K$, forward reaction occurs to make Q larger. If $Q > K$, reverse reaction occurs to make Q smaller. If $Q = K$, system is already at equilibrium, so nothing happens.

$2HI(g) \rightleftharpoons H_2(g) + I_2(g) \quad K = 0.016$

Suppose $P\ HI = 0.100$ atm, no H_2 or I_2

$Q = 0$, system shifts to right

Suppose $P\ HI = P\ H_2 = P\ I_2 = 0.100$ atm

$Q = 1$, system shifts to left

B. Extent of reaction

1. Start with pure HI at 0.100 atm. Equilibrium pressures of all species?

$2HI(g) \rightleftharpoons H_2(g) + I_2(g) \quad K = 0.016$

	2HI(g)	H_2(g)	I_2(g)
P_o	0.100	0.000	0.000
ΔP	-2x	+x	+x
P_{eq}	0.100 - 2x	x	x

$$\frac{x^2}{(0.100 - 2x)^2} = 0.016 \qquad \frac{x}{0.100 - 2x} = 0.13 \qquad x = 0.010$$

P HI = 0.080 atm; P H_2 = P I_2 = 0.010 atm

2. Start with pure N_2O_4 at 0.100 atm. Equilibrium partial pressures of N_2O_4, NO_2?

$N_2O_4(g) \rightleftharpoons 2NO_2(g) \qquad K = 11$

	N_2O_4	NO_2
P_o	1.00	0.00
ΔP	-x	+2x
P_{eq}	1.00 - x	2x

$4x^2/(1.00 - x) = 11$; $x = 0.78$ (by quadratic formula)

P NO_2 = 1.56 atm P N_2O_4 = 0.22 atm

LECTURE 2½

I Effect of Changes in Conditions on Equilibrium Position

Le Chatelier's principle: If system at equilibrium is subjected to stress, reaction occurs in a direction so as to partially relieve the stress.

A. Adding or removing a gaseous species

$2HI(g) \rightleftharpoons H_2(g) + I_2(g)$

If H_2 is added, part of it reacts; P HI greater than before equilibrium was disturbed, P I_2 less. P H_2 is intermediate between original equilibrium value and that immediately after equilibrium was disturbed.

If H_2 is removed, some HI decomposes to bring P H_2 back part way to its original value. P I_2 increases, P HI decreases.

B. Increasing volume

Immediate effect is to lower concentration of molecules. To counteract this, reaction occurs which increases number of molecules in gas phase.

	Expansion	Contraction
$N_2O_4(g) \rightleftharpoons 2NO_2(g)$	→	←
$N_2(g) + 3H_2(g) \rightleftharpoons 2NH_3(g)$	←	→
$H_2(g) + I_2(g) \rightleftharpoons 2HI(g)$	no effect	

C. Change in temperature

Effect of increase in T is partially counteracted if endothermic process occurs. Hence, endothermic process is favored by increase in T.

	ΔH	increase T	decrease T
$N_2O_4(g) \rightleftharpoons 2NO_2(g)$	+58.2 kJ	→	←
$N_2(g) + 3H_2(g) \rightleftharpoons 2NH_3(g)$	-92.4 kJ	←	→

If forward reaction is endothermic, K increases as T increases; if exothermic, K decreases as T increases.

Temperature dependence of K: $\ln K_2/K_1 = \Delta H[1/T_1 - 1/T_2]/R$

DEMONSTRATIONS

1. Mechanical model of chemical equilibrium: J. Chem. Educ. 67 598 (1990)
2. N_2O_4 - NO_2 system: Test. Dem. 19, 131, 167; Shak. 2 180
3. Le Chatelier's principle: Test. Dem. 19, 221; J. Chem. Educ. 47 A735 (1970)
4. Ostwald process: Test. Dem. 38, 99, 169, 224
5. Catalytic oxidation of ammonia: Shak. 2 214
6. Air pollution: J. Chem. Educ. 64 893 (1987)

QUIZZES

Quiz 1

1. Consider the equilibrium: $2NO(g) \rightleftharpoons N_2(g) + O_2(g)$ $\Delta H = -181kJ$
   a. Starting with pure NO at a partial pressure of 0.80 atm at a certain temperature, its equilibrium partial pressure is 0.60 atm. Calculate K.
   b. Which way does the equilibrium shift if the system is compressed?
   c. Which way does the equilibrium shift if the temperature is increased?

Quiz 2

1. Consider the equilibrium

   $PCl_5(g) \rightleftharpoons PCl_3(g) + Cl_2(g)$ K = 0.050

a. Starting with pure PCl_5, its equilibrium partial pressure is found to be 0.10 atm. What are the equilibrium partial pressures of PCl_3 and Cl_2?

b. Some PCl_3 is added to the equilibrium system in (a). How will the equilibrium partial pressures of all three species compare to their values before equilibrium was disturbed?

Quiz 3

1. For the equilibrium: $2HI(g) \rightleftharpoons H_2(g) + I_2(g)$

   K = 0.016 at 520°C

   If you start with pure HI at a partial pressure of 1.00 atm, what will be the equilibrium partial pressures of all species at 520°C?

Quiz 4

1. Consider the system: $2NO_2(g) \rightleftharpoons 2NO(g) + O_2(g)$

   a. Starting with pure NO_2 at 1.00 atm, the equilibrium partial pressure of O_2 is found to be 0.20 atm. Calculate K.

   b. If you start with all three gases at identical partial pressures of 1.00 atm, which way will the system move to reach equilibrium?

   c. Calculate K for: $NO(g) + \frac{1}{2} O_2(g) \rightleftharpoons NO_2(g)$

Quiz 5

1. Consider the system:

   $2NO(g) + O_2(g) \rightleftharpoons 2NO_2(g) \quad \Delta H = -114$ kJ

   Starting with NO at 2.00 atm and O_2 at 1.00 atm at 500°C, the equilibrium partial pressure of NO_2 is 1.00 atm. Calculate

   a. K at 500°C b. K at 400°C

Answers

Quiz 1 1. a. 0.028 b. no effect c. to the left

Quiz 2 1. a. 0.071 atm b. PCl_3 greater, Cl_2 less, PCl_5 greater

Quiz 3 1. P HI = 0.80 atm; P H_2 = P I_2 = 0.10 atm

Quiz 4 1. a. 0.089 b. to the left c. 3.4

Quiz 5 1. a. 2.0 b. 0.14

PROBLEMS

1. 75 s

3.

0.200	0.150	0.125	0.110	0.100	0.100
0.250	0.150	0.100	0.070	0.050	0.050
0.000	0.050	0.075	0.090	0.100	0.100

5. a. $\dfrac{(P\ CH_4) \times (P\ H_2S)^2}{(P\ CS_2) \times (P\ H_2)^4}$ b. $\dfrac{(P\ H_2O)^2 \times P\ O_2}{(P\ H_2O_2)^2}$

c. $\dfrac{(P\ CO_2)^3 \times (P\ H_2O)^4}{(P\ C_3H_8) \times (P\ O_2)^5}$

7. a. $P\ CO_2/P\ CO$ b. $P\ O_2$ c. $(P\ SO_2)^2/(P\ O_2)^3$

9. a. $C_3H_6O(l) \rightleftharpoons C_3H_6O(g)$ $K = P\ C_3H_6O$

b. $7H_2(g) + 2NO_2(g) \rightleftharpoons 2NH_3(g) + 4H_2O(g)$

$K = \dfrac{(P\ H_2O)^4 \times (P\ NH_3)^2}{(P\ NO_2)^2 \times (P\ H_2)^7}$

c. $3Cl_2(g) + CS_2(l) \rightleftharpoons CCl_4(g) + S_2Cl_2(l)$ $K = 1/(P\ Cl_2)^3$

11. a. $2\ O_3(g) \rightleftharpoons 3\ O_2(g)$

b. $2SO_2(g) + O_2(g) \rightleftharpoons 2SO_3(g)$

c. $N_2(g) + 3H_2(g) \rightleftharpoons 2NH_3(g)$

d. $4HCl(g) + O_2(g) \rightleftharpoons 2H_2O(g) + 2Cl_2(g)$

e. $ZnO(s) + H_2(g) \rightleftharpoons Zn(s) + H_2O(g)$

13. a. $K = (2.2 \times 10^{-3})^2 = 4.8 \times 10^{-6}$

b. $K = 1/(2.2 \times 10^{-3})^4 = 4.3 \times 10^{10}$

15. $K = 21 \times (0.034)^2 = 0.024$

17. $K = [(8 \times 10^1)/(1 \times 10^{30})]^{\frac{1}{2}} = 9 \times 10^{-15}$

19. a. $CO_2(g) + C(s) \rightleftharpoons 2CO(g)$

b. $K = (0.43)^2/0.092 = 2.0$

21. $P = MRT$; $RT = 0.0821\ \dfrac{L\cdot atm}{mol\cdot K} \times 1273\ K = 105\ \dfrac{atm}{mol/L}$

$P\ CO = 2.1 \times 10^{-4}$ atm, $P\ O_2 = 1.0 \times 10^{-4}$ atm, $P\ CO_2 = 26$ atm

$K = \dfrac{(2.1 \times 10^{-4})^2 \times (1.0 \times 10^{-4})}{(26)^2} = 6.5 \times 10^{-15}$

23. $P\ NOBr = 0.624$ atm; $P\ NO = 1.577$ atm $- 0.624$ atm $= 0.953$ atm

$P\ Br_2 = 0.427$ atm $- 0.312$ atm $= 0.115$ atm

$$K = \frac{(0.624)^2}{(0.953)^2 \times 0.115} = 3.73$$

25. a. $Q = 0.30 \times 0.16/0.50 = 0.096$; no; $Q > K$

b. to the left

27. a. $Q = (0.10) \times (0.10)^2/(0.10)^2 = 0.10 < K$ to the right

b. $Q = 0 < K$ to the right

c. $Q = (0.010) \times (0.040)^2/(0.20)^2 = 4.0 \times 10^{-4} < K$ to right

29. a

31. $\frac{(P\ NO)^2 \times 0.127}{(0.384)^2} = 1.74 \times 10^{-4}$; $P\ NO = 0.0142$ atm

33. $1.3 \times 10^5 = \frac{(P\ H_2S)^2}{(0.0200)^2 \times (0.0400)}$ $P\ H_2S = 1.4$ atm

35. $4.0 \times P\ Cl_2 = 3.33$; $P\ Cl_2 = 0.83$ atm

37. $SO_2(g) + NO_2(g) \rightleftharpoons NO(g) + SO_3(g)$

	$SO_2(g)$	$NO_2(g)$	$NO(g)$	$SO_3(g)$
P_o	0.075	0.075	0.000	0.000
ΔP	- x	- x	+ x	+ x
P_{eq}	0.075 - x	0.075 - x	x	x

$$\frac{x^2}{(0.075 - x)^2} = 85.0 \quad \frac{x}{0.075 - x} = 9.2 \text{ ; } x = 0.068$$

$P\ NO = P\ SO_3 = 0.068$ atm; $P\ SO_2 = P\ NO_2 = 0.007$ atm

39. $H_2(g) + C_2N_2(g) \rightleftharpoons 2HCN(g)$

	$H_2(g)$	$C_2N_2(g)$	$2HCN(g)$
P_o	0.200	0.150	0.0100
ΔP	- x	- x	+ 2x
P_{eq}	0.200 - x	0.150 - x	0.0100 + 2x

$$1.50 = \frac{(0.0100 + 2x)^2}{(0.200 - x) \times (0.150 - x)}$$

Solving by the quadratic formula, $x = 0.0623$ atm

$P\ H_2 = 0.138$ atm; $P\ C_2N_2 = 0.088$ atm; $P\ HCN = 0.135$ atm

41. a. $K = \frac{(0.012) \times (0.012)}{(0.010) \times (0.020)} = 0.72$

b.

	$CO(g)$	+	$H_2O(g)$	$\rightleftharpoons$	$CO_2(g)$	+	$H_2(g)$
P_o	0.020		0.020		0.012		0.012
ΔP	$-x$		$-x$		$+x$		$+x$
P_{eq}	$0.020 - x$		$0.020 - x$		$0.012 + x$		$0.012+x$

$$0.72 = \frac{(0.012 + x)^2}{(0.020 - x)^2} \qquad 0.85 = \frac{0.012 + x}{0.020 - x} \qquad x = 0.003$$

$P\ CO = P\ H_2O = 0.017$ atm; $P\ CO_2 = P\ H_2 = 0.015$ atm

43.

	$N_2(g)$	+	$O_2(g)$	$\rightleftharpoons$	$2NO(g)$
P_o	0.35		0.35		0
ΔP	$-x$		$-x$		$+2x$
P_{eq}	$0.35 - x$		$0.35 - x$		$2x$

$$\frac{(2x)^2}{(0.35 - x)^2} = 2.5 \times 10^{-3} \ ; \qquad \frac{2x}{0.35 - x} = 0.050$$

$$x = 0.0085$$

$P\ NO = 0.017$ atm; $P\ N_2 = P\ O_2 = 0.34$ atm

45. a. R b. R c. no effect d. R e. F

47. a. to the left b. to the right c. no effect

49. a or c

51. $\ln 2 = \Delta H°[1/298 - 1/308]/8.31 \qquad \Delta H° = 53$ kJ

53. At equilibrium, $P\ SO_2 = 0.16$ atm, $P\ O_2 = 0.37$ atm

$$K = \frac{0.49}{(0.16) \times (0.37)^{\frac{1}{2}}} = 5.0$$

55. exothermic; low T, high P

57. a. $SO_3(g) + H_2O(l) \longrightarrow H_2SO_4(l)$

b. burning coal, roasting sulfides

c. $SO_2(g) + \frac{1}{2}\ O_2(g) \longrightarrow SO_3(g)$

59. See p. 349

61. $P = MRT$

At 700°C, $P = 79.9M$; $P\ CO = 4.0$ atm, $P\ CO_2 = 8.0$ atm

$$K = (4.0)^2/8.0 = 2.0$$

At 600°C, P = 71.7M; P_{CO} = 71.7(0.01)atm = 0.7 atm

P_{CO_2} = 71.7(0.12)atm = 8.6 atm

$K = (0.7)^2/8.6 = 0.06$

63. $2.1 = x^2/(0.27 - x)$; solving by the quadratic formula, x = 0.24

$P_{tot} = 0.27 + x = 0.51$ atm

65. $$P\ O_2(\text{orig}) = \frac{(1.00\ \text{mol})(0.0821\ \text{L·atm/mol·K})(1800\ \text{K})}{5.0\ \text{L}} = 30\ \text{atm}$$

	$O_2(g)$ ⇌	$2\ O(g)$
P_o	30	0
P	- x	+ 2x
P_{eq}	30 - x	2x

$4x^2/(30 - x) = 1.7 \times 10^{-8}$ $\quad x^2 \approx 1.7 \times 10^{-8} \times 30/4$

$x = 3.6 \times 10^{-4}$ atm

$$\text{mass O} = \frac{\mathcal{M}PV}{RT} = 3.9 \times 10^{-4}\ \text{g}$$

mass % $= (3.9 \times 10^{-4}/32.0) \times 100 = 1.2 \times 10^{-3}$

$$\text{no. O atoms} = 3.9 \times 10^{-4}\ \text{g} \times \frac{6.022 \times 10^{23}\ \text{atoms}}{16.00\ \text{g}}$$

$= 1.5 \times 10^{19}$ atoms

66. $aA(g) + bB(g) \rightleftharpoons cC(g) + dD(g)$

$P_A = [A]RT$; $(P_A)^a = [A]^a(RT)^a$

$$K = \frac{(P_C)^c \times (P_D)^d}{(P_A)^a \times (P_B)^b} = \frac{[C]^c \times [D]^d}{[A]^a \times [B]^b} \times (RT)^{(c+d-a-b)}$$

$K = K_c(RT)^{\Delta n_g}$

67. $$n\ I_2 = 0.0370\ \text{L} \times 0.200\ \frac{\text{mol}}{\text{L}} \times \frac{1}{2} = 0.00370\ \text{mol}$$

$n\ H_2 = 0.00370$ mol

$$n\ HI = \frac{3.20}{127.9}\ \text{mol} - 0.00740\ \text{mol} = 0.0176\ \text{mol}$$

$$K = \frac{P\,I_2 \times P\,H_2}{(P\,HI)^2} = \frac{n\,I_2 \times n\,H_2}{(n\,HI)^2} \times \frac{(RT/V)^2}{(RT/V)^2} = \frac{n\,I_2 \times n\,H_2}{(n\,HI)^2}$$

$$K = \frac{(0.00370) \times (0.00370)}{(0.0176)^2} = 0.0442$$

68. $0.45 = \frac{(x/2)^{\frac{1}{2}}(x)}{1.00 - x}$ By trial and error, $x = 0.48$

$P\,SO_3 = 0.52$ atm

69. At equilibrium: $P\,XeF_4 = 0.10$ atm, $P\,Xe = 0.10$ atm
$P\,F_2 = 0.20$ atm

$$K = \frac{0.10}{(0.10)(0.20)^2} = 25$$

new equilibrium pressures: $P\,XeF_4 = 0.15$ atm, $P\,Xe = 0.05$ atm

$$25 = \frac{0.15}{(0.05)(P\,F_2)^2} \qquad P\,F_2 = 0.35 \text{ atm}$$

$P_o = 0.35$ atm + 2(0.15 atm) = 0.65 atm

70. Let $P_o\,I_2 = 1.00$ atm

$P_{tot} = 1.40 = 1.00 - x + 2x$; $x = 0.40$ atm

At equilibrium: $P\,I_2 = 0.60$ atm, $P\,I = 0.80$ atm

$K = (0.80)^2/0.60 = 1.1$

71. a. $P_oC_6H_5CH_2OH = \frac{(1.50/108.13 \text{ mol})(0.0821 \text{ L·atm/mol·K})(523 \text{ K})}{2.0 \text{ L}}$

$= 0.30$ atm

$x^2/(0.30 - x) = 0.56$; $x = 0.22$ $P = 0.22$ atm

b. P_{eq} benzyl alcohol = 0.08 atm

$\text{mass} = \frac{(108.13 \text{ g/mol})(0.08 \text{ atm})(2.0 \text{ L})}{(0.0821 \text{ L·atm/mol·K})(523 \text{ K})} = 0.4$ g

CHAPTER 13
Acids and Bases

LECTURE NOTES

This is a chapter that should be covered slowly and carefully. Students have trouble with it because there are so many different concepts. Curiously, they do better with equilibrium calculations than with qualitative concepts such as

- predicting whether a given species (in particular a salt) will give an acidic, basic, or neutral solution.

- writing a chemical equation to show why a species is acidic or basic.

To master these skills, students must know which acids and bases are strong (Chapter 4).

You will probably want to spend 2½ lectures, perhaps 3, on this chapter.

LECTURE 1

I Bronsted-Lowry Model of Acids and Bases

Acid is proton donor, base is proton acceptor

$$HB(aq) + A^-(aq) \rightleftharpoons HA(aq) + B^-(aq)$$

HB, HA are acids; A^-, B^- are bases. HB - B^- and HA - A^- are conjugate acid-base pairs.

II Acidic and Basic water Solutions

A. In any aqueous solution, there is an equilibrium between H_3O^+ (H^+) ions and OH^- ions.

$$H_2O \rightleftharpoons H^+(aq) + OH^-(aq)$$

$$K_w = [H^+] \times [OH^-] = 1.0 \times 10^{-14} \text{ at } 25°C$$

1. Pure water: $[H^+] = [OH^-] = 1.0 \times 10^{-7}$ M; neutral solution
2. Acidic solution: $[H^+] > 1.0 \times 10^{-7} M > [OH^-]$

3. Basic solution: $[OH^-] > 1.0 \times 10^{-7}$ M $> [H^+]$

In seawater, $[H^+] = 5 \times 10^{-9}$ M; $[OH^-] = ?$

$[OH^-] = (1.0 \times 10^{-14})/(5 \times 10^{-9}) = 2 \times 10^{-6}$ M

b. $pH = -\log_{10}[H^+]$

Neutral solution: pH = 7.0
Acidic solution: pH < 7.0
Basic solution: pH > 7.0

Suppose $[H^+] = 2.4 \times 10^{-6}$ M; calculate pH

$$pH = -\log_{10}(2.4 \times 10^{-6}) = 5.62$$

Suppose pH = 8.68; calculate $[H^+]$

Use 10^x or INV and LOG keys $[H^+] = 10^{-8.68} = 2.1 \times 10^{-9}$ M

$pOH = -\log_{10}[OH^-]$; $pOH + pH = 14.00$ at 25°C

III Weak Acids

A. Molecules

$$HF(aq) + H_2O \rightleftharpoons H_3O^+(aq) + F^-(aq)$$

B. Cations

$$NH_4^+(aq) + H_2O \rightleftharpoons H_3O^+(aq) + NH_3(aq)$$

$$Zn(H_2O)_4^{2+}(aq) + H_2O \rightleftharpoons H_3O^+(aq) + Zn(H_2O)_3(OH)^+(aq)$$

LECTURE 2

I Weak Acids

A. Dissociation constants

$$HB(aq) \rightleftharpoons H^+(aq) + B^-(aq) \qquad K_a = \frac{[H^+] \times [B^-]}{[HB]}$$

1. Calculation of K_a:

pH of 0.100 M $HC_2H_3O_2$ solution is 2.87; $K_a = ?$

$[H^+] = [C_2H_3O_2^-] = 1.3 \times 10^{-3}$ M

$[HC_2H_3O_2] = 0.100$ M $- 0.0013$ M $= 0.099$ M

$$K_a = \frac{(1.3 \times 10^{-3})^2}{0.099} = 1.7 \times 10^{-5}$$

2. Calculation of $[H^+]$ in solution

a. Find $[H^+]$ in 0.200 M $HC_2H_3O_2$ ($K_a = 1.8 \times 10^{-5}$)

Let $[H^+] = x$; then $[C_2H_3O_2^-] = x$, $[HC_2H_3O_2] = 0.200 - x$

$$\frac{x^2}{0.200 - x} = 1.8 \times 10^{-5}$$

To solve, assume $0.200 - x = 0.200$, in which case:

$$x^2 = 0.200(1.8 \times 10^{-5}) = 3.6 \times 10^{-6}$$

$$x = (3.6 \times 10^{-6})^{\frac{1}{2}} = 1.9 \times 10^{-3} \text{ M}$$

Approximation is generally O.K. if $x < 5\%$ of original concentration, i.e., % ionization < 5%

In this case, % ionization $= \dfrac{1.9 \times 10^{-3}}{0.200} \times 100 = 1.0\%$

b. Find $[H^+]$ in 0.100 M HF ($K_a = 6.9 \times 10^{-4}$)

$$\frac{x^2}{0.100 - x} = 6.9 \times 10^{-4}$$

Set $0.100 - x = 0.100$, solve: $x = 8.3 \times 10^{-3} > 5\%$

Make second approximation:

$$\frac{x^2}{0.100 - 0.008} = 6.9 \times 10^{-4}; \; x = 8.0 \times 10^{-3} \text{ M}$$

II Weak Bases

A. Molecules

$$NH_3(aq) + H_2O \rightleftharpoons OH^-(aq) + NH_4^+(aq)$$

B. Anions derived from weak acids

$$F^-(aq) + H_2O \rightleftharpoons HF(aq) + OH^-(aq)$$

LECTURE 3

I Weak Bases

A. Dissociation constant

1. Expression for K_b:

$$NH_3: \; K_b = \frac{[NH_4^+] \times [OH^-]}{[NH_3]}$$

2. Relation between K_a and K_b

$K_a \times K_b = K_w$

	K_a		K_b
HF	6.9×10^{-4}	F^-	1.4×10^{-11}
HAc	1.8×10^{-5}	Ac^-	5.6×10^{-10}
NH_4^+	5.6×10^{-10}	NH_3	1.8×10^{-5}

Strength of base is inversely related to that of conjugate weak acid.

3. Calculation of $[OH^-]$ in solution of weak base

   pH of 0.10 M NaF solution?

   $$\frac{[HF] \times [OH^-]}{[F^-]} = 1.4 \times 10^{-11} \; ; \quad \frac{[OH^-]^2}{0.10} = 1.4 \times 10^{-11}$$

   $[OH^-] = 1.2 \times 10^{-6}$ M; pH = 8.08

II Acid-Base Properties of Salt Solutions

A. Cations

1. Spectator - derived from strong bases:

   Li^+, Na^+, K^+, Ca^{2+}, Sr^{2+}, Ba^{2+}

2. Acidic - all other cations, including those of the transition metals.

B. Anions

1. Spectator - derived from strong acids

   Cl^-, Br^-, I^-, NO_3^-, ClO_4^-, SO_4^{2-}

2. Basic - anions derived from weak acids, such as F^-, NO_2-

C. Overall results

Salt	Cation	Anion	
$NaNO_3$	Na^+(S)	NO_3^-(S)	neutral
KF	K^+(S)	F^-(B)	basic
$FeCl_2$	Fe^{2+}(A)	Cl^-(S)	acidic

If cation is acidic, anion basic, compare K_a, K_b values:

NH_4F: $K_a = 5.6 \times 10^{-10}$, $K_b = 1.4 \times 10^{-11}$; acidic

DEMONSTRATIONS

1. pH of household chemicals: Shak. 3
*2. Fountain effects with NH_3, HCl: Shak. 3 92
3. Indicator colors: Test. Dem. 12, 61, 147; J. Chem. Educ. 61 172 (1984); Shak. 3 33, 41
4. Grape juice as an indicator: Test. Dem. 204
*5. Cabbage juice as an indicator: J. Chem. Educ. 69 66 (1992); Shak. 3 50
6. Acid-base properties of salts: Test. Dem. 62; Shak. 3 103
7. pH of phosphate salts: Test. Dem. 40
8. Acidic properties of CO_2 in water: Shak. 3 114

* See also Shakhashiri Videotapes, Demonstrations 1 - 4

QUIZZES

Quiz 1

1. The pH of a solution is 2.68. Calculate $[H^+]$, $[OH^-]$, and pOH.
2. Classify solutions of the following salts as acidic, basic, or neutral. Where appropriate, write balanced net ionic equations to explain your answers.

   a. $NaNO_3$ b. $ZnCl_2$ c. KCN

Quiz 2

1. Calculate the pH of 0.200 M $HC_2H_3O_2$ ($K_a = 1.8 \times 10^{-5}$). What is $[OH^-]$ in this solution?
2. Write balanced net ionic equations to explain why

   a. a solution of $Fe(NO_3)_3$ is acidic

   b. a solution of Na_2CO_3 is basic

   c. H_2O can act as a Bronsted-Lowry acid

Quiz 3

1. Given that K_a of HNO_2 is 6.0×10^{-4}, calculate

   a. K_b of the NO_2^- ion

   b. $[OH^-]$ in 0.100 M $NaNO_2$

2. Give the formula of

   a. the conjugate acid of HCO_3^-

   b. the conjugate base of HCO_3^-

   c. an amphiprotic anion

Quiz 4

1. The pH of a 0.100 M solution of a certain weak acid is 3.68. Calculate K_a.

2. Classify the following as weak acids, strong acids, weak bases, or strong bases.

   a. HF   b. CO_3^{2-} c. NH_3 d. $HClO_4$ e. $Cu(H_2O)_4^{2+}$

Quiz 5

1. Given that K_b of NH_3 is 1.8×10^{-5}, calculate

   a. K_a for NH_4^+

   b. the pH of a 1.00 M NH_4Cl solution

2. Write balanced net ionic equations to explain why

   a. a water solution of Na_2CO_3 is basic

   b. a water solution of HClO is acidic

   c. a water solution of NH_4F is acidic

Answers

Quiz 1 1. 2.1×10^{-3} M, 4.8×10^{-12} M, 11.32

2. a. neutral

   b. acidic: $Zn(H_2O)_4^{2+}(aq) + H_2O \rightleftharpoons Zn(H_2O)_3(OH)^+(aq) + H_3O^+(aq)$

   c. basic: $CN^-(aq) + H_2O \rightleftharpoons OH^-(aq) + HCN(aq)$

Quiz 2 1. 2.72, 5.3×10^{-12} M

2. a. $Fe(H_2O)_6^{3+}(aq) + H_2O \rightleftharpoons Fe(H_2O)_5(OH)^{2+}(aq) + H_3O^+(aq)$

   b, c. $CO_3^{2-}(aq) + H_2O \rightleftharpoons OH^-(aq) + HCO_3^-(aq)$

Quiz 3 1. a. 1.7×10^{-11} b. 1.3×10^{-6} M

2. a. H_2CO_3 b. CO_3^{2-} c. HCO_3^-

Quiz 4 1. 4.4×10^{-7} 2. a. WA   b. WB   c. WB   d. SA   e. WA

Quiz 5 1. a. 5.6×10^{-10} b. 4.63

2 a. $CO_3^{2-}(aq) + H_2O \rightleftharpoons OH^-(aq) + HCO_3^-(aq)$

b. $HClO(aq) + H_2O \rightleftharpoons H_3O^+(aq) + ClO^-(aq)$

c. $NH_4^+(aq) + H_2O \rightleftharpoons H_3O^+(aq) + NH_3(aq)$

PROBLEMS

1. a. BA: H_3O^+, H_2SO_3 BB: HSO_3^-, H_2O

   CP: H_3O^+, H_2O; H_2SO_3, HSO_3^-

   b. BA: HF, H_2O BB: OH^-, F^-

   CP: HF, F^-; H_2O, OH^-

   c. BA: NH_4^+, H_3O^+ BB: H_2O, NH_3

   CP: NH_4^+, NH_3; H_3O^+, H_2O

3. a. BA b. BB c. BB (BA in principle)

5. a. H_3O^+ b. H_2CO_3 c. $CH_3NH_3^+$ d. HS^- e. $Fe(H_2O)_6^{2+}$

7. a. 2.0; acidic b. 4.00; acidic c. 4.40; acidic d. 9.21; basic

9.

	a.	b.	c.	d.
$[H^+]$	1×10^{-4} M	3.0×10^{-9} M	1.0 M	2.5×10^{-13}
$[OH^-]$	1×10^{-10} M	3.3×10^{-6} M	1.0×10^{-14} M	4.0×10^{-2}

11. Solution B: $[OH^-] = 1.4 \times 10^{-6}$ M

    A is more basic; B has lower pH

13. 2.6, 3.8

15. a. 3×10^{-11} M b. $3 \times 10^{-2}/(3 \times 10^{-11}) = 1 \times 10^9$

17. a. $[H^+] = 0.30$ M, pH = 0.52

    b. $[H^+] = 6.00 \text{ M} \times \frac{10.0}{300} = 0.200$ M; pH = 0.699

19. $n\ H^+ = 0.245 \text{ L} \times 0.0235 \text{ mol/L} + 0.438 \text{ L} \times 0.554 \text{ mol/L}$
    $= 0.248$ mol

    $[H^+] = 0.248 \text{ mol}/0.683 \text{ L} = 0.363$ M; pH = 0.440

21. a. $[OH^-] = 0.50$ M; $[H^+] = 2.0 \times 10^{-14}$ M; pH = 0.30

b. $[OH^-] = \frac{100.0/40.00 \text{ mol}}{0.500 \text{ L}} = 5.00$ M; $[H^+] = 2.00 \times 10^{-15}$ M

pOH = -0.699

23. $[OH^-] = \frac{32.1/40.00 \text{ mol} + 56.3/56.11 \text{ mol}}{1.75 \text{ L}} = 1.03$ M

$[H^+] = 9.7 \times 10^{-15}$ M; pH = 14.01

25. a. $Ni(H_2O)_5(OH)^+(aq) + H_2O \rightleftharpoons H_3O^+(aq) + Ni(H_2O)_4(OH)_2(aq)$

b. $Al(H_2O)_6{}^{3+}(aq) + H_2O \rightleftharpoons H_3O^+(aq) + Al(H_2O)_5(OH)^{2+}(aq)$

c. $H_2S(aq) + H_2O \rightleftharpoons H_3O^+(aq) + HS^-(aq)$

d. $H_2PO_4{}^-(aq) + H_2O \rightleftharpoons H_3O^+(aq) + HPO_4^{2-}(aq)$

e. $Cr(H_2O)_5(OH)^{2+}(aq) + H_2O \rightleftharpoons H_3O^+(aq) + Cr(H_2O)_4(OH)_2{}^+(aq)$

f. $HClO_2(aq) + H_2O \rightleftharpoons H_3O^+(aq) + ClO_2{}^-(aq)$

27. a. $HSO_3{}^-(aq) \rightleftharpoons H^+(aq) + SO_3{}^{2-}(aq) \quad K_a = \frac{[H^+] \times [SO_3{}^{2-}]}{[HSO_3{}^-]}$

b. $HPO_4{}^{2-}(aq) \rightleftharpoons H^+(aq) + PO_4{}^{3-}(aq) \quad K_a = \frac{[H^+] \times [PO_4{}^{3-}]}{[HPO_4{}^{2-}]}$

c. $HNO_2(aq) \rightleftharpoons H^+(aq) + NO_2{}^-(aq) \quad K_a = \frac{[H^+] \times [NO_2{}^-]}{[HNO_2]}$

29. 6.36, 10.33

31. a. B < A < C < D b. D

33. $K_a = \frac{(8.1 \times 10^{-4})^2}{0.029} = 2.3 \times 10^{-5}$

35. $[HC_7H_5O_2]_o = \frac{1.00/122.12 \text{ mol}}{0.350 \text{ L}} = 0.0234$ M

$[H^+] = 1.2 \times 10^{-3}$ M

$K_a = \frac{(1.2 \times 10^{-3})^2}{0.0222} = 6.5 \times 10^{-5}$

37. a. $[H^+]^2 \approx 3.8 \times 10^{-4}$; $[H^+] = 1.9 \times 10^{-2}$ M

b. $[H^+]^2 \approx 0.33 \times 1.9 \times 10^{-4}$; $[H^+] = 7.9 \times 10^{-3}$ M

39. a. $[H^+]^2 \approx (1.3)(1.4 \times 10^{-4})$; $[H^+] = 1.3 \times 10^{-2}$ M

b. $[OH^-] = 7.7 \times 10^{-13}$ M c. pH = 1.89

d. % diss. = $\frac{1.3 \times 10^{-2}}{1.3} \times 100\% = 1.0\%$

41. $[H^+]^2 \approx 0.20 \times 1.4 \times 10^{-3}$; $[H^+] \approx 0.017$ M

[HA] = 0.20 M - 0.017 M = 0.18 M

$[H^+]^2 = 0.18 \times 1.4 \times 10^{-3}$; $[H^+] = 0.016$ M; pH = 1.80

43. $[H^+]^2 = (0.10)(4.4 \times 10^{-7})$; $[H^+] = 2.1 \times 10^{-4}$ M; pH = 3.68

45. $[HCO_3^-] = [H^+] = 2.1 \times 10^{-4}$ M

$[CO_3^{2-}] = K_2 = 4.7 \times 10^{-11}$ M

47. a. $NH_3(aq) + H_2O \rightleftharpoons NH_4^+(aq) + OH^-(aq)$

b. $NO_2^-(aq) + H_2O \rightleftharpoons HNO_2(aq) + OH^-(aq)$

c. $C_6H_5NH_2(aq) + H_2O \rightleftharpoons C_6H_5NH_3^+(aq) + OH^-(aq)$

d. $CO_3^{2-}(aq) + H_2O \rightleftharpoons HCO_3^-(aq) + OH^-(aq)$

e. $F^-(aq) + H_2O \rightleftharpoons HF(aq) + OH^-(aq)$

f. $HCO_3^-(aq) + H_2O \rightleftharpoons H_2CO_3(aQ) + OH^-(aq)$

49. b > a > d > c

51. a, c, e

53. a. 1.9×10^{-11} b. 2.6×10^{-5}

55. $C_7H_5O_2^-(aq) + H_2O \rightleftharpoons HC_7H_5O_2(aq) + OH^-(aq)$

a. 1.5×10^{-10}

b. $[OH^-]^2 \approx 0.23(1.5 \times 10^{-10})$; $[OH^-] = 5.9 \times 10^{-6}$ M; pH = 8.77

57. $[OH^-]^2 \approx 0.30(1.8 \times 10^{-5})$; $[OH^-] = 2.3 \times 10^{-3}$ M; pH = 11.36

59. a. A b. B c. B d. N e. B f. B

61. a. $NH_4^+(aq) + H_2O \rightleftharpoons H_3O^+(aq) + NH_3(aq)$

b. $CN^-(aq) + H_2O \rightleftharpoons HCN(aq) + OH^-(aq)$

c. $PO_4^{3-}(aq) + H_2O \rightleftharpoons HPO_4^{2-}(aq) + OH^-(aq)$

e. $HCO_3^-(aQ) + H_2O \rightleftharpoons H_2CO_3(aq) + OH^-(aq)$

f. $F^-(aq) + H_2O \rightleftharpoons HF(aq) + OH^-(aq)$

63. $HBr < ZnCl_2 < BaI_2 < NH_3 < KOH$

65. a. $FeCl_3$, $Fe(NO_3)_3$, $FeBr_3$, $Fe(ClO_4)_3$

b. $NaNO_3$, KNO_3, $LiNO_3$, $Ba(NO_3)_2$

c. $Ba(ClO)_2$, $Ba(C_2H_3O_2)_2$, $Ba(HCO_3)_2$, $Ba(CN)_2$

d. CsCl, CsBr, CsI, $CsNO_3$

67. a. ascorbic acid b. See p. 379 c. 60 mg

69. a. See p. 378 b. acetic < lactic < citric < oxalic

71. a. HCl, HBr, - - b. NH_3, Na_2CO_3, - -

c. NH_4Cl, $CuSO_4$, - - d. Na_2CO_3, KF, - -

e. NaCl, KNO_3, - --

73. greater

75. $[H^+]^2 \approx 0.025 \times 1.2 \times 10^{-5}$; $[H^+] = 5.5 \times 10^{-4}$ M; pH = 3.26

76.
$$H_2CO_3(aq) \rightleftharpoons H^+(aq) + HCO_3^-(aq) \quad K_{1a}$$
$$HCO_3^-(aq) \rightleftharpoons H^+(aq) + CO_3^{2-}(aq) \quad K_{2a}$$
$$H_2CO_3(aq) \rightleftharpoons 2H^+(aq) + CO_3^{2-}(aq) \quad K = K_{1a}K_{2a} = 2.1 \times 10^{-17}$$

77. Measure pH of 0.10 M $AgNO_3$; if neutral, AgOH is a strong base

78. a. $H^+(aq) + OH^-(aq) \longrightarrow H_2O \quad \Delta H° = -55.8$ kJ

$$\Delta H = 1.00 \text{ L} \times 0.100 \frac{\text{mol}}{\text{L}} \times \frac{-55.8 \text{ kJ}}{1 \text{ mol}} = -5.58 \text{ kJ}$$

b. $HF(aq) + OH^-(aq) \longrightarrow H_2O + F^-(aq)$

$\Delta H° = \Delta H_f° F^- + \Delta H_f° H_2O - \Delta H_f° OH^- - \Delta H_f° HF = -68.3$ kJ

$\Delta H = -6.83$ kJ

79. % diss. = $\frac{[H^+]}{[HA]_o}$; $[H^+]^2 = K_a \times [HA]_o$; $[H^+] = K_a^{\frac{1}{2}} \times [HA]_o^{\frac{1}{2}}$

% diss. = $K_a^{\frac{1}{2}}/[HA]_o^{\frac{1}{2}}$

80. Consider 1000 g solution

n $HC_2H_3O_2$ = 50.0/60.05 mol = 0.833 mol; mass water = 950 g

molality before dissociation = 0.833 mol/0.950 kg = 0.877 m

V solution = 1000 g/1.006 g/mL = 994 mL

$[HC_2H_3O_2]$ = 0.833 mol/0.994 L = 0.838 M

$[H^+] = (0.838 \times 1.8 \times 10^{-5})^{\frac{1}{2}} = 3.9 \times 10^{-3}$ M $\approx$ molality H^+

total molality = 0.0039 + 0.0039 + (0.877 - 0.0039) = 0.881 m

$\Delta T_f = 1.86 \frac{°C}{m} \times 0.881\ m = 1.64°C \qquad T_f = -1.64°C$

CHAPTER 14
Acid-Base and Precipitation Equilibria

LECTURE NOTES

This chapter focuses upon the application of equilibrium principles to acid-base titrations, buffers, and precipitation reactions. The most difficult topic is that of buffers. Students have a great deal of trouble calculating the pH of a buffer after addition of strong acid or base. Such a calculation requires that they apply two different concepts: stoichiometry and equilibrium principles.

Students generally find Section 14.3 straightforward. They have fewer difficulties with K_{sp} calculations than with any other type of equilibrium constant. Overall, 2 - 2½ lectures should be sufficient for this chapter, depending upon how much time you wish to spend on acid-base titrations.

LECTURE 1

I Acid-Base Titrations

A. Strong acid - strong base

$$H^+(aq) + OH^-(aq) \longrightarrow H_2O \qquad K = 1/K_w = 1.0 \times 10^{14}$$

pH = 7 at end point, changes rapidly near end point (Fig. 14.1)

B. Weak acid-strong base

$$HA(aq) + OH^-(aq) \longrightarrow A^-(aq) + H_2O \qquad K = 1/K_b$$

pH > 7 at end point, changes slowly near end point (Fig 14.2)

C. Strong acid - weak base

$$H^+(aq) + A^-(aq) \longrightarrow HA(aq) \qquad K = 1/K_a$$

pH < 7 at end point, changes slowly near end point (Fig. 14.3)

D. Choice of indicator

$$HIn(aq) \rightleftharpoons H^+(aq) + In^-(aq)$$

HIn and In^- have different colors; color of solution depends upon ratio of concentrations.

$$\frac{[HIn]}{[In^-]} = \frac{[H^+]}{K_a}$$

End point occurs when $[H^+] = K_a$. About pH 5 for methyl red ($K_a = 10^{-5}$), 7 for bromthymol blue ($K_a = 10^{-7}$), 9 for phenolphthalein ($K_a = 10^{-9}$).

II <u>Buffers</u>

A. Prepared by adding both the weak acid HB and its conjugate base B^- to water.

$$[H^+] = K_a \times \frac{[HB]}{[B^-]} = K_a \times \frac{n\ HB}{n\ B^-}$$

Calculate $[H^+]$ in a solution prepared by adding 0.200 mol of HAc, 0.200 mol NaAc to one liter. Neglect the small amount of acetic acid that dissociates.

$$[H^+] = 1.8 \times 10^{-5} \times \frac{0.200}{0.200} = 1.8 \times 10^{-5}\ M \qquad pH = 4.74$$

B. Effect of adding strong acid or base to buffer

$$HB(aq) + OH^-(aq) \longrightarrow H_2O + B^-(aq)$$

$$B^-(aq) + H^+(aq) \longrightarrow HB(aq)$$

As a result of these reactions, H^+ or OH^- ions are consumed, so do not drastically change the pH.

Calculate pH of buffer in (A) after addition of 0.020 mol NaOH.

	orig. conc.	change	final conc.
HAc	0.200	- 0.020	0.180
Ac^-	0.200	+ 0.020	0.220

$$[H^+] = 1.8 \times 10^{-5} \times \frac{0.180}{0.220} = 1.5 \times 10^{-5}$$

pH = 4.82 (slightly greater than originally)

LECTURE 2

I <u>Buffers</u>

A. <u>Choice of system</u>. Note that since HB and B^- are present in roughly equal amounts, $[H^+]$ of the buffer is roughly equal to K_a of the weak acid. To establish a buffer of pH 7, choose a system such as $H_2PO_4^-$, HPO_4^{2-}, where $K_a = 6.2 \times 10^{-8}$

At pH 7.00, $\frac{[H_2PO_4^-]}{[HPO_4^{2-}]} = \frac{1.0 \times 10^{-7}}{6.2 \times 10^{-8}} = 1.6$

Add 160 mL of 0.100 M $H_2PO_4^-$ to 100 mL of 0.100 M HPO_4^{2-}

II <u>Solubility Equilibrium</u>; K_{sp}

A. <u>Expression for</u> K_{sp}

$AgCl(s) \rightleftharpoons Ag^+(aq) + Cl^-(aq)$

$K_{sp} = [Ag^+] \times [Cl^-]$

$PbCl_2(s) \rightleftharpoons Pb^{2+}(aq) + 2Cl^-(aq)$

$K_{sp} = [Pb^{2+}] \times [Cl^-]^2 = 1.7 \times 10^{-5}$

B. <u>Uses of</u> K_{sp}

1. Calculation of concentration of one ion, knowing that of other.

What is $[Pb^{2+}]$ in a solution in equilibrium with $PbCl_2$ ($K_{sp} = 1.7 \times 10^{-5}$) if $[Cl^-] = 0.020$ M?

$[Pb^{2+}] = \frac{1.7 \times 10^{-5}}{(2.0 \times 10^{-2})^2} = 0.042$ M

2. Determination of whether precipitate will form.

If $P < K_{sp}$, no precipitate (equilibrium not established)

If $P > K_{sp}$, precipitate forms until P becomes equal to K_{sp}

Suppose enough Ag^+ is added to a solution 0.001 M in CrO_4^{2-} to make conc. $Ag^+ = 0.001$ M. Will Ag_2CrO_4 precipitate? ($K_{sp} = 2 \times 10^{-12}$)

P = (orig. conc. $Ag^+)^2$ x (orig. conc. CrO_4^{2-})

$= (1 \times 10^{-3})^2 \times (1 \times 10^{-3}) = 1 \times 10^{-9} > K_{sp}$

precipitate forms

3. Determination of solubility

a. Pure water

$PbCl_2(s) \rightleftharpoons Pb^{2+}(aq) + 2Cl^-(aq)$

$[Pb^{2+}] = s$; $[Cl^-] = 2s$

$4s^3 = K_{sp} = 1.7 \times 10^{-5}$; $s = 1.6 \times 10^{-2}$ M

b. In solution containing a common ion.

Solubility $PbCl_2$ in 0.100 M HCl?

$[Pb^{2+}] = s \quad [Cl^-] = s + 0.10 \approx 0.10$

$s(0.10)^2 = 1.7 \times 10^{-5}$; $s = 1.7 \times 10^{-3}$ M

Solubility about one tenth that in pure water

DEMONSTRATIONS

1. Acid-base titrations: J. Chem. Educ. 56 194 (1979)
2. Heat of neutralization: Shak. 1 15
3. Carbon dioxide and limewater: Shak. 1 329
4. Amphoteric properties of metal hydroxides: Shak. 3 128
5. Titration curves: Shak. 3 167
6. Buffers: Test. Dem. 62, 128, 155, 210; J. Chem. Educ. 60 493 (1983), 62 337, 436, 608 (1985); Shak. 3 173
7. Buffering action of Alka-Seltzer: Shak. 3 186
8. Solubility product constant: Test. Dem. 131, 173
9. Selective precipitation: J. Chem. Educ. 65 359 (1988)
10. Equilibrium between silver chloride and silver chromate: J. Chem. Educ. 54 618 (1977); 65 621 (1988)
11. Common ion effect: Test. Dem. 19, 86

QUIZZES

Quiz 1

1. A buffer contains 0.100 mol NH_4^+ and 0.050 mol NH_3. Taking K_a $NH_4^+ = 5.6 \times 10^{-10}$, calculate

   a. the pH of the buffer

   b. the pH of the buffer after addition of 0.010 mol HCl

2. Explain briefly why a NH_4^+ - NH_3 buffer would not be suitable at pH 7.

Quiz 2

Consider the titration of $HC_2H_3O_2$ with NaOH.

1. Write an equation for the acid-base reaction.

2. Calculate K for the reaction.

3. The pH at the equivalence point of the titration is about 8.5. What is the ratio $[HIn]/[In^-]$ at this point if K_a HIn = 1.0×10^{-8}?

Quiz 3

1. Taking K_{sp} of $PbCl_2$ to be 1.7×10^{-5}, determine whether a precipitate will form when 50.00 mL of 0.10 M $Pb(NO_3)_2$ is mixed with 50.00 mL of 0.020 M HCl.

2. Write the expression relating the water solubility s (moles per liter) of Ag_2CrO_4 to K_{sp}.

Quiz 4

1. What volume of 0.100 M $HC_2H_3O_2$ has to be added to 25.0 mL of 0.100 M $NaC_2H_3O_2$ to form a buffer with a pH of 5.00? Take K_a $HC_2H_3O_2$ = 1.8×10^{-5}.

2. What kind of indicator would you use for the titration of $NaC_2H_3O_2$ with HCl? Explain.

Quiz 5

1. Write the expression for K_{sp} of Bi_2S_3.

2. Taking K_{sp} of Bi_2S_3 to be 1×10^{-99}, calculate its solubility (moles per liter) in

   a. water b. 0.100 M Na_2S

Answers

Quiz 1 1. a. 8.95 b. 8.81
2. Would need very high ratio of $[NH_4^+]$ to $[NH_3]$; would have very little capacity for acid

Quiz 2 1. $HC_2H_3O_2(aq) + OH^-(aq) \longrightarrow C_2H_3O_2^-(aq) + H_2O$
2. 1.8×10^9 3. 0.3

Quiz 3 1. no; $5 \times 10^{-6} < 1.7 \times 10^{-5}$ 2. $K_{sp} = 4s^3$

Quiz 4 1. 13.9 mL
2. Indicator that changes color at about pH 5; solution is weakly acidic because of $HC_2H_3O_2$ formed.

Quiz 5 1. $K_{sp} = [Bi^{3+}]^2 \times [S^{2-}]^3$
2. a. 6×10^{-21} M b. 1×10^{-48} M

PROBLEMS

1. a. $H^+(aq) + OH^-(aq) \longrightarrow H_2O$

   b. $H^+(aq) + NH_3(aq) \longrightarrow NH_4^+(aq)$

   c. $HF(aq) + CN^-(aq) \longrightarrow HCN(aq) + F^-(aq)$

   d. $HNO_2(aq) + OH^-(aq) \longrightarrow H_2O + NO_2^-(aq)$

3. a. $H^+(aq) + CHO_2^-(aq) \longrightarrow HCHO_2(aq)$

   b. $H^+(aq) + OH^-(aq) \longrightarrow H_2O$

   c. $H^+(aq) + NH_3(aq) \longrightarrow NH_4^+(aq)$

5. a. $K = 1/K_w = 1.0 \times 10^{14}$

   b. $K = 1/K_a\ NH_4^+ = 1/(5.6 \times 10^{-10}) = 1.8 \times 10^9$

   c. $K = K_aHF/K_aHCN = \dfrac{6.9 \times 10^{-4}}{5.8 \times 10^{-10}} = 1.2 \times 10^6$

7. a. $K = 1/K_a\ HCHO_2 = 1/(1.9 \times 10^{-4}) = 5.3 \times 10^3$

   b. $K = 1/K_w = 1.0 \times 10^{14}$

   c. $K = 1/K_a\ NH_4^+ = 1/(5.6 \times 10^{-10}) = 1.8 \times 10^9$

9. a. Ph b. MO c. MO d. any

11. $n\ OH^- = 3.500 \times 10^{-2}\ L \times 0.2500\ mol/L = 8.750 \times 10^{-3}\ mol$

   a. $n\ H^+ = 1.000 \times 10^{-2}\ L \times 0.4375\ mol/L = 4.375 \times 10^{-3}\ mol$

   $n\ OH^-$ excess $= 4.375 \times 10^{-3}\ mol$

   $[OH^-] = \dfrac{4.375 \times 10^{-3}\ mol}{4.500 \times 10^{-2}\ L} = 0.09722\ M$; pH = 12.99

   b. $n\ H^+ = 8.750 \times 10^{-3}\ mol$; pH = 7.00

   c. $n\ H^+ = 3(4.375 \times 10^{-3}\ mol) = 13.125 \times 10^{-3}\ mol$

   $n\ H^+$ excess $= 4.375 \times 10^{-3}\ mol$

   $[H^+] = \dfrac{4.375 \times 10^{-3}\ mol}{6.500 \times 10^{-2}\ L} = 0.06731\ M$ pH = 1.172

13. a. K^+, NO_2^- b. basic

15. a. $[H^+]^2 = (0.100)(1.9 \times 10^{-4})$; $[H^+] = 4.4 \times 10^{-3}$ M; pH = 2.36

b. 2.50×10^{-3} mol

c. $HCHO_2(aq) + OH^-(aq) \longrightarrow CHO_2^-(aq) + H_2O$

d. $[CHO_2^-] = \dfrac{2.50 \times 10^{-3}\ \text{mol}}{4.17 \times 10^{-2}\ \text{L}} = 0.0600$ M

$[OH^-]^2 = (0.0600)(5.3 \times 10^{-11}) = 3.2 \times 10^{-12}$

$[OH^-] = 1.8 \times 10^{-6}$ M; pOH = 5.74; pH = 9.26

17. a. $[H^+] = 5.6 \times 10^{-10} \times \dfrac{[NH_4^+]}{[NH_3]} = 2.2 \times 10^{-10}$ M; pH = 9.66

b. $[H^+] = 5.6 \times 10^{-10}$ M; pH = 9.25

c. $[H^+] = 1.1 \times 10^{-9}$ M; pH = 8.96

d. $[H^+] = 1.1 \times 10^{-8}$ M; pH = 7.96

19. n NO_2^- = 0.020 mol

n HNO_2 = 0.250 L x 0.040 mol/L = 0.010 mol

$[H^+] = 6.0 \times 10^{-4} \times \dfrac{0.010}{0.020} = 3.0 \times 10^{-4}$ M; pH = 3.52

21. a. lactic acid; lactate   b. $H_2PO_4^-$; HPO_4^{2-} c. NH_4^+; NH_3

23. a. $1.0 \times 10^{-3} = 1.9 \times 10^{-4} \times R$; R = 5.3

b. $1.0 \times 10^{-3} = 1.9 \times 10^{-4} \times \dfrac{n}{0.200}$; n = 1.1 mol

c. $1.0 \times 10^{-3} = 1.9 \times 10^{-4} \times \dfrac{0.100}{m/68.01}$ m = 1.3 g

25. $[H^+] = 1.8 \times 10^{-5} \times \dfrac{12.50/60.05}{15.00/82.03} = 2.0 \times 10^{-5}$ M; 4.70, 4.70

27. $[H^+] = 5.6 \times 10^{-10} \times n\ NH_4^+/(0.250 \times 0.150\ \text{mol}) = 1.0 \times 10^{-9}$

n NH_4^+ = 0.067 mol

m = 0.067 mol x 80.05 g/mol = 5.4 g

29. $[H^+] = 1.8 \times 10^{-3}$ M; $K_a = \dfrac{(1.8 \times 10^{-3})^2}{0.048} = 6.8 \times 10^{-5}$

$[H^+] = 6.8 \times 10^{-5} \times \frac{0.050}{0.035} = 9.7 \times 10^{-5}$; pH = 4.01

31. a. $[H^+] = 5.6 \times 10^{-10}$ M; pH = 9.25

    b. $[H^+] = 5.6 \times 10^{-10} \times \frac{0.21}{0.19} = 6.2 \times 10^{-10}$; pH = 9.21

    c. $[H^+] = 5.6 \times 10^{-10} \times \frac{0.18}{0.22} = 4.6 \times 10^{-10}$; pH = 9.34

33. a. $[H^+] = 4.7 \times 10^{-11} \times 4.0 = 1.9 \times 10^{-10}$ M; pH = 9.72

    b. $\frac{4.0 \times 10^{-10}}{4.7 \times 10^{-11}} = R = 8.5$

35. b

37. $n\ HC_3H_5O_2 = 2.00\ g \times \frac{1\ mol}{74.08\ g} = 0.0270\ mol$

    $n\ NaOH = 0.45\ g \times \frac{1\ mol}{40.00\ g} = 0.0112\ mol$

    after reaction: $n\ HC_3H_5O_2 = 0.0158\ mol$

    $n\ C_3H_5O_2^- = 0.0112\ mol$

    $[H^+] = 1.4 \times 10^{-5} \times \frac{0.0158}{0.0112} = 2.0 \times 10^{-5}$ M; pH = 4.70

39. a. $Zr(OH)_4(s) \rightleftharpoons Zr^{2+}(aq) + 4\ OH^-(aq)$; $[Zr^{4+}] \times [OH^-]^4$

    b. $PbBr_2(s) \rightleftharpoons Pb^{2+}(aq) + 2Br^-(aq)$; $[Pb^{2+}] \times [Br^-]^2$

    c. $K_2SiF_6(s) \rightleftharpoons 2K^+(aq) + SiF_6^{2-}(aq)$; $[K^+]^2 \times [SiF_6^{2-}]$

    d. $Bi_2S_3(s) \rightleftharpoons 2Bi^{3+}(aq) + 3S^{2-}(aq)$; $[Bi^{3+}]^2 \times [S^{2-}]^3$

41. a. $TlI(s) \rightleftharpoons Tl^+(aq) + I^-(aq)$

    b. $Eu(OH)_3(s) \rightleftharpoons Eu^{3+}(aq) + 3\ OH^-(aq)$

    c. $Pb_3(PO_4)_2(s) \rightleftharpoons 3Pb^{2+}(aq) + 2PO_4^{3-}(aq)$

    d. $Zn(OH)_2(s) \rightleftharpoons Zn^{2+}(aq) + 2\ OH^-(aq)$

43.

	$[Mg^{2+}]$	$[OH^-]$
a.	6×10^{-4} M	1×10^{-4} M
b.	6×10^{-3} M	3×10^{-5} M
c.	9×10^{-4} M	8×10^{-5} M
d.	2×10^{-5} M	5×10^{-4} M

45. a. $[Ag^+]^3 = \frac{1 \times 10^{-16}}{1.0 \times 10^{-4}}$ $[Ag^+] = 1 \times 10^{-4}$ M

b. $[Ca^{2+}]^3 = \frac{1 \times 10^{-33}}{(1.0 \times 10^{-4})^2}$ $[Ca^{2+}] = 5 \times 10^{-9}$ M

c. $[Al^{3+}] = \frac{1 \times 10^{-20}}{1.0 \times 10^{-4}} = 1 \times 10^{-16}$ M

47. a. $[Ag^+] = \frac{1.9 \times 10^{-3}}{1.00 \times 10^{-2}} = 0.19$ M

b. $[C_2H_3O_2^-] = \frac{1.9 \times 10^{-3}}{0.500} = 3.8 \times 10^{-3}$

% left $= \frac{3.8 \times 10^{-3}}{1.00 \times 10^{-2}} \times 100\% = 38\%$

49. $P = (1 \times 10^{-4})(5 \times 10^{-3}) = 5 \times 10^{-7} > K_{sp}$; yes

51. $[Fe^{2+}] = 0.100 \text{ M} \times \frac{25.0}{75.0} = 0.0333$ M

$[OH^-] = 1.0 \times 10^{-5} \text{ M} \times \frac{50.0}{75.0} = 6.7 \times 10^{-6}$ M

$P = (3.33 \times 10^{-2})(6.7 \times 10^{-6})^2 = 1.5 \times 10^{-12} < K_{sp}$; no

53. $s = \frac{1.07 \times 10^{-3} \text{ g}}{1 \text{ L}} \times \frac{1 \text{ mol}}{114.95 \text{ g}} = 9.31 \times 10^{-6}$ mol/L

$K_{sp} = (9.31 \times 10^{-6})^2 = 8.67 \times 10^{-11}$

55. $Mg_3(PO_4)_2(s) \rightleftharpoons 3Mg^{2+}(aq) + 2PO_4^{3-}(aq)$

molar mass $Mg_3(PO_4)_2$ = 262.84 g/mol

a. $(3s)^3(2s)^2 = 108s^5 = 1 \times 10^{-24}$

$s = 6 \times 10^{-6} \text{ M} = 2 \times 10^{-3}$ g/L

b. $[PO_4^{3-}]^2 = \frac{1 \times 10^{-24}}{(1.0 \times 10^{-2})^3}$ $[PO_4^{3-}] = 1 \times 10^{-9}$ M

$s = [PO_4^{3-}]/2 = 5 \times 10^{-10} \text{ M} = 1 \times 10^{-7}$ g/L

c. $[Mg^{2+}]^3 = \frac{1 \times 10^{-24}}{(2.0 \times 10^{-2})^2}$ $[Mg^{2+}] = 1.4 \times 10^{-7}$ M

$s = 1.4 \times 10^{-7} \text{ M}/3 = 5 \times 10^{-8} \text{ M} = 1 \times 10^{-5}$ g/L

57. $CO_2(g) + H_2O \rightleftharpoons H_2CO_3(aq) \rightleftharpoons H^+(aq) + HCO_3^-(aq)$

addition of H^+ ions shifts equilibrium to left, forming CO_2

59. a. $4.0 \times 10^{-8} = 4.4 \times 10^{-7} \times R$; $R = 0.091$

b. $[H^+] = 4.4 \times 10^{-7} \times \frac{0.191}{0.900} = 9.3 \times 10^{-8}$ M; pH = 7.03

c. $[H^+] = 4.4 \times 10^{-7} \times \frac{0.082}{1.009} = 3.6 \times 10^{-8}$ M; pH = 7.44

61. a. benzoate ion is basic

b. HNO_2 is a weak acid, only slightly dissociated

c. reacts with both H^+, OH^-

63. a. $\leftarrow$ b. $\rightarrow$ c. $\rightarrow$ d. $\leftarrow$

65. $K = \frac{K_{sp}\ Ag_2CrO_4}{(K_{sp}AgCl)^2} = \frac{1 \times 10^{-12}}{(1.8 \times 10^{-10})^2} = 3 \times 10^7$; yes

67. a. $[H^+]^2 = 1.8 \times 10^{-5}$; $[H^+] = 4.2 \times 10^{-3}$ M; pH = 2.37

b. $[H^+] = 1.8 \times 10^{-5}$ M; pH = 4.74

c. $[H^+] = 1.8 \times 10^{-5} \times \frac{0.10}{49.90} = 3.6 \times 10^{-8}$ M; pH = 7.44

d. 0.500 M $NaC_2H_3O_2$

$[OH^-]^2 = 5.6 \times 10^{-10} \times 0.50$; $[OH^-] = 1.7 \times 10^{-5}$ M; pH = 9.22

e. $[OH^-] = 1.000\ M \times \frac{0.10}{100.10} = 1.0 \times 10^{-3}$ M; pH = 11.00

f. $[OH^-] = 1.000\ M \times \frac{50.00}{150.00} = 0.333$ M; pH = 13.52

68. $Mg(OH)_2(s) \rightleftharpoons Mg^{2+}(aq) + 2\ OH^-(aq)$; $K_1 = K_{sp}Mg(OH)_2$

$2NH_4^+(aq) \rightleftharpoons 2H^+(aq) + 2NH_3(aq)$; $K_2 = (K_a NH_4^+)^2$

$2H^+(aq) + 2\ OH^-(aq) \rightleftharpoons 2H_2O$; $K_3 = 1/K_w^2$

$$K = \frac{(6 \times 10^{-12})(5.6 \times 10^{-10})^2}{(1.0 \times 10^{-14})^2} = 2 \times 10^{-2}$$

$4s^3 = (0.02)(0.20)^2$; $s = 0.06$ M

water: $4s^3 = 6 \times 10^{-12}$; $s = 1 \times 10^{-4}$ M

69. $1.0 \times 10^{-5} = 1.8 \times 10^{-5} \times \frac{[HC_2H_3O_2]}{[C_2H_3O_2^-]}$

$\frac{[HC_2H_3O_2]}{[C_2H_3O_2^-]} = 0.56$; fraction neutralized $= \frac{1.00}{1.56} = 0.64$

32 mL NaOH

70. $K = \frac{K_{sp}\ CaF_2}{(K_a HF)^2} = \frac{1.5 \times 10^{-10}}{(6.9 \times 10^{-4})^2} = 3.2 \times 10^{-4}$

$[H^+] = 1.9 \times 10^{-4} \times 0.30/0.20 = 2.8 \times 10^{-4}$ M

$4s^3/(2.8 \times 10^{-4})^2 = 3.2 \times 10^{-4}$; $s = 1.8 \times 10^{-4}$ M

71. $[Ag^+] = (1.6 \times 10^{-10})/(2.0 \times 10^{-2}) = 8.0 \times 10^{-9}$ M

$[I^-] = (1 \times 10^{-16})/(8.0 \times 10^{-9}) = 1 \times 10^{-8}$ M

72. a. $[OH^-]^2 = \frac{6 \times 10^{-12}}{5.6 \times 10^{-2}} = 1 \times 10^{-10}$ $[OH^-] = 1 \times 10^{-5}$ M

b. $Al(OH)_3$: $P = (4 \times 10^{-7})(1 \times 10^{-5})^3 = 4 \times 10^{-22} > K_{sp}$

$Fe(OH)_3$: $P = (2 \times 10^{-7})(1 \times 10^{-5})^3 = 2 \times 10^{-22} > K_{sp}$

$Al(OH)_3$, $Fe(OH)_3$ precipitate

c. $[OH^-]^2 = \frac{6 \times 10^{-12}}{2.8 \times 10^{-2}}$ $[OH^-] = 1 \times 10^{-5}$ M

$[Al^{3+}] = \frac{2 \times 10^{-31}}{1 \times 10^{-15}} = 2 \times 10^{-16}$ M; $[Fe^{3+}] = 3 \times 10^{-24}$ M

virtually all

d. 0.028 mol $Mg(OH)_2$ + 4×10^{-7} mol $Al(OH)_3$ + 2×10^{-7} mol $Fe(OH)_3$

0.028 mol $\times \frac{58.32\ g}{1\ mol} = 1.6$ g

73. $\log_{10}[H^+] = \log_{10}K_a + \log_{10}\frac{[HB]}{[B^-]}$

multiply by -1: $pH = pK_a + \log_{10}\frac{[B^-]}{[HB]}$

CHAPTER 15
Complex Ions; Coordination Compounds

LECTURE NOTES

One thing to keep in mind throughout this chapter is the amount of "jargon" involved. Such terms as "chelating agent", "coordination number", "square planar" and "ligand" are unfamiliar to students. They also have trouble visualizing the structure of a complex given only its formula, e.g. $Co(en)_2(NH_3)_2^{3+}$.

This is a short chapter with two main topics:

- geometry of complexes, including isomerism. Here, it helps to emphasize the geometric relationships in an octahedron, a figure which is unfamiliar to many students.

- electronic structure of complexes. Here, we discuss only the crystal field model and restrict it to octahedral complexes. Before doing that, we review briefly the electronic structure of transition metal ions, covered way back in Chapter 6.

Two lectures should be quite adequate; it can be covered in 1½ lectures if you delete Section 15.2 on nomenclature.

LECTURE 1

I Complex Ions and Coordination Compounds

A. A complex ion consists of a central metal cation (usually derived from a transition metal) joined by coordinate covalent bonds to two or more molecules or anions called ligands.

Complex ion	Cation	Ligands	Coord. No.
$Ag(NH_3)_2^+$	Ag^+	2 NH_3 molecules	2
$Cu(H_2O)_4^{2+}$	Cu^{2+}	4 H_2O molecules	4
$Fe(CN)_6^{3-}$	Fe^{3+}	6 CN^- ions	6

Reaction between cation and ligand can be considered to be an acid-base reaction:

$Ag^+(aq) + 2NH_3(aq) \rightarrow Ag(NH_3)_2^+(aq)$

Lewis acid   Lewis base

(Lewis acid accepts electron pair, Lewis base donates it)

Coordination compound contains complex ion. Examples:

$[Cu(H_2O]_4]SO_4$ $[Ag(NH_3)_2]NO_3$ $K_3[Fe(CN)_6]$

Charge of central ion in $Zn(H_2O)_3(OH)^+$?

$+1 = -1 + x; x = +2$

B. Nature of ligands; ordinarily contain at least one unshared pair of electrons.

$H - \ddot{N} - H$ (with H below N) $H - \ddot{O} - H$ (two lone pairs) $(:\ddot{O} - H)^-$ $(:\ddot{Cl}:)^-$

If the ligand contains two or more unshared pairs on different, nonadjacent atoms, it can act as a chelating agent, forming more than one bond with the central metal atom.

$H - \ddot{N} - CH_2 - CH_2 - \ddot{N} - H$ (with H below each N)   ethylenediamine (en)

forms complexes such as $Cu(en)_2^{2+}$, $Cr(en)_3^{3+}$

II <u>Nomenclature of Complex Ions</u>

A. Cation $Co(NH_3)_4Cl_2^+$ tetraamminedichlorocobalt(III)

$Ni(en)_3^{2+}$ tris(ethylenediamine)nickel(II)

1. Identify each ligand, naming in alphabetical order
2. Give number of each ligand (di, tri, - - or bis, tris - -)
3. Give oxidation number of central metal

B. Anion $Zn(OH)_4^{2-}$ tetrahydroxozincate(II)

III Geometry of Complex Ions

A. Coordination no. = 2: linear, 180° bond angle

$(NH_3 - Ag - NH_3)^+$

B. Coordination no. = 4

1. Tetrahedral $Zn(NH_3)_4^{2+}$

2. Square planar $Cu(NH_3)_4^{2+}$

C. Coordination number = 6

Octahedral: $Co(NH_3)_6^{3+}$

LECTURE 2

I Geometric Isomerism

A. Square planar

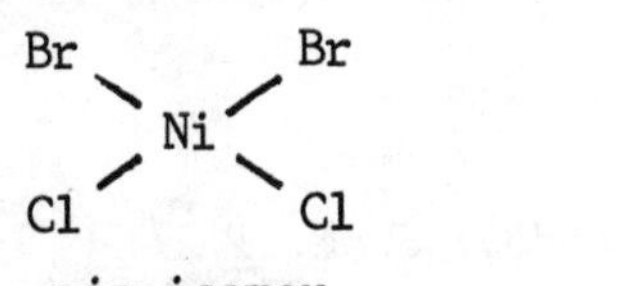

cis isomer
(like groups close)

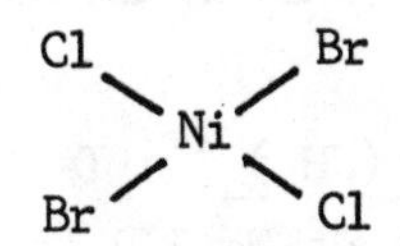

trans isomer
(like groups far apart)

B. Octahedral: $Co(NH_3)_4Cl_2^+$

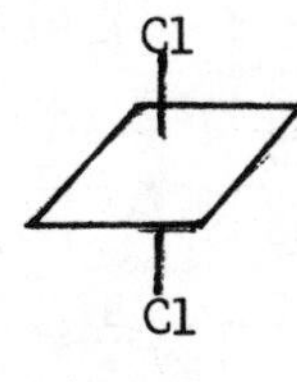

trans

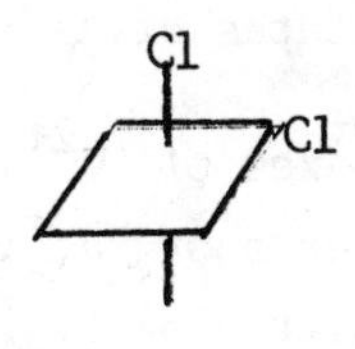

cis

II Electronic Structure Transition Metal Ions

No outer s electrons

Cr^{3+} [Ar] $3d^3$ Co^{2+} [Ar] $3d^7$ Zn^{2+} [Ar] $3d^{10}$

III Crystal Field Model

Approach of six ligands to transition metal cation splits d orbitals into two sets of different energy:

high energy: $d_{x^2-y^2}$, d_{z^2}

low energy: d_{xy}, d_{yz}, d_{xz}

A. Splitting energy Δ_o small: Hund's rule is followed, giving "high-spin" complex with the maximum number of unpaired electrons

Co^{2+} (↑)(↑) 3 unpaired electrons
(↑↓)(↑↓)(↑)

B. Splitting energy Δ_o large: electrons fill lower-energy orbitals, giving "low-spin" complex.

Co^{2+} (↑)( ) 1 unpaired electron

(↑↓)(↑↓)(↑↓)

C. Color Caused by electron transitions between two sets of d orbitals.

DEMONSTRATIONS

1. Formation of $Cu(NH_3)_4^{2+}$: Test. Dem. 22

*2. Iodo complexes of Hg(II) (orange tornado): Shak. 1 271

3. Complexes of Co^{2+}: Test. Dem. 222; J. Chem. Educ. 57 453 (1980); Shak. 1 280

4. Complexes of Cu^{2+}: J. Chem. Educ. 62 798 (1985); Shak. 1 314

*5. Complexes and precipitates of Ag^{+}: J. Chem. Educ. 57 813 (1980); Shak. 1 307

6. Precipitates and complexes of Cu^{2+}: Shak. 1 318

7. Precipitates and complexes of Fe^{3+}: Shak. 1 338

*8. Complexes and precipitates of Ni^{2+}: J. Chem. Educ. 57 900 (1980); Shak. 1 299

9. Complexes and precipitates of Pb^{2+}: Shak. 1 286

10. EDTA complexes: Test. Dem. 132

* See also Shakhashiri Videotapes, Demonstrations 12, 13, 14

QUIZZES

Quiz 1

1. Show the geometry, including isomers, of

   a. $AgCl_2^-$ b. $Zn(en)_2^{2+}$ (tetrahedral) c. $Co(en)_2Cl_2^+$

2. Give the formula of a simple salt (e.g., NaCl) which would be similar to each of the following in conductivity and colligative properties.

   a. $K_2[Zn(OH)_4]$ b. $[Co(NH_3)_3(en)Cl]SO_4$ c. $[Pt(NH_3)_4]Cl_2$

Quiz 2

1. Draw orbital diagrams for low-spin and high-spin complexes of Co^{2+} and Co^{3+}.

2. Give the charge and the coordination no. of the metal in:

   a. $[Co(NH_3)_4(en)]Cl_2$ b. $[Pt(en)(H_2O)Cl_3]^+$ c. $Al(OH)_6{}^{3-}$

Quiz 3

1. Give the coordination number of the central metal atom in:

   a. $Cu(en)_2^+$ b. $Co(NH_3)_3Cl_3$ c. $Fe(ox)_3^{3-}$

2. Show the geometric isomers of $[Pt(NH_3)(H_2O)Cl_2]$ (square planar)

Quiz 4

1. Draw electron configurations for the following transition metal ions:

   a. Mn^{2+} b. Fe^{2+} c. Ni^{2+}

2. Draw orbital diagrams for low-spin octahedral complexes of the cations in (1).

Quiz 5

1. Draw all of the possible structures for the following complexes

   a. $Cr(NH_3)_3(NO_3)_3$ b. $Co(en)_2Br_2^+$ c. $Ni(NH_3)_5(H_2O)^{2+}$

2. What is the coordination number and oxidation number of the central metal in each of the complexes in Question 1?

Answers

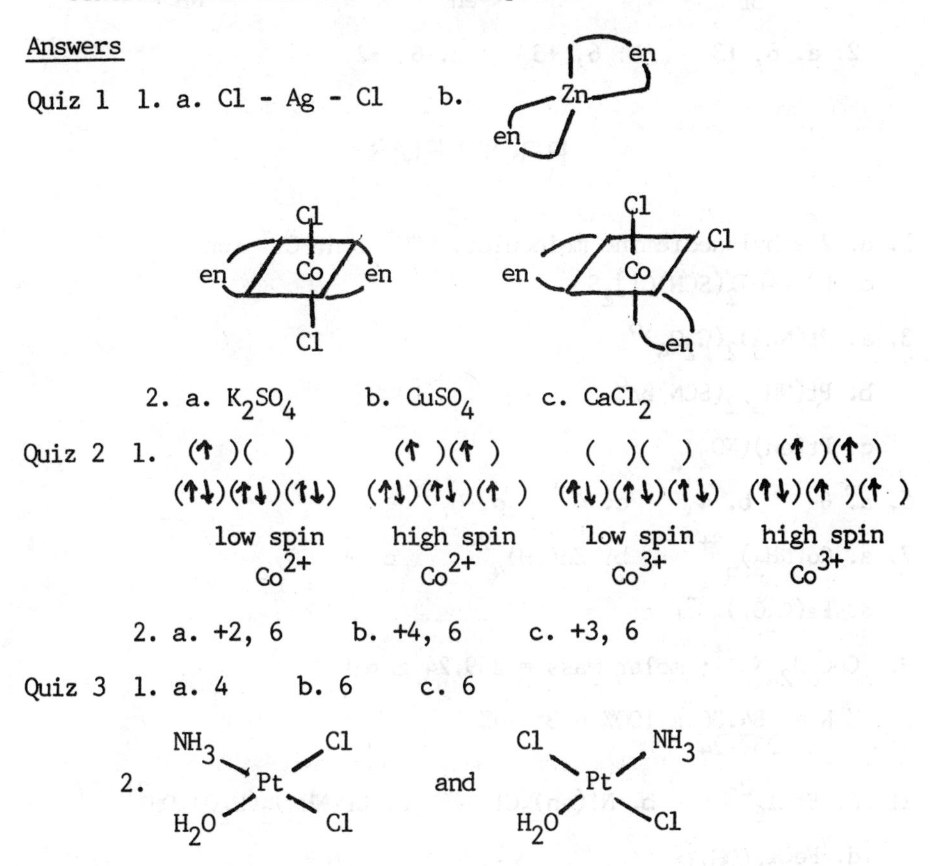

Quiz 4 1. a. [Ar]$3d^5$ b. [Ar]$3d^6$ c. [Ar]$3d^8$

2. a. ( )( ) b. ( )( ) c. (↑)(↑)

(↑↓)(↑↓)(↑) (↑↓)(↑↓)(↑↓) (↑↓)(↑↓)(↑↓)

Quiz 5

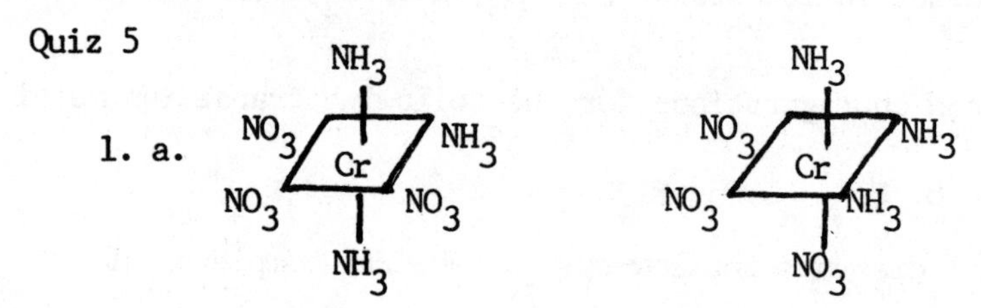

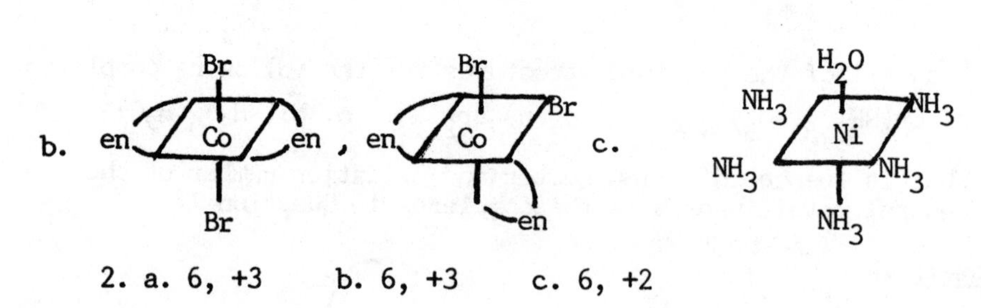

2. a. 6, +3 b. 6, +3 c. 6, +2

PROBLEMS

1. a. 2 ethylenediamine molecules, SCN^- ion, Cl^- ion b. +3
   c. $[Co(en)_2(SCN)Cl]_2S$

3. a. $Pt(NH_3)_2(C_2O_4)^o$
   b. $Pt(NH_3)_2(SCN)Br^o$
   c. $Pt(en)(NO_2)_2{}^o$

5. a. 6 b. 4 c. 4 d. 6

7. a. $Co(NH_3)_6{}^{3+}$ b. $Zn(OH)_4{}^{2-}$ c. $Ag(CN)_2{}^-$
   d. $Fe(C_2O_4)_3{}^{3-}$

9. $CoC_6H_{24}N_6{}^{3+}$; molar mass = 239.24 g/mol

$$\% N = \frac{84.06}{239.24} \times 100\% = 35.14\%$$

11. a. $PtCl_6{}^{2-}$ b. $Ni(en)_2Cl_2$ c. $Cr(NH_3)_2(H_2O)_3OH^{2+}$
   d. $FeCl_5(OH)^{3-}$

13. a. Pentaamminechlororuthenium(III)
    b. Tris(ethylenediamine)manganese(II)
    c. Tetrachloroplatinate(II)
    d. Pentaammineiodochromium(III)

15. a. Diamminebis(ethylenediamine)copper(II)
    b. Tetracyanonickelate(II)
    c. Bis(ethylenediamine)zinc(II)
    d. Hexachlorovanadate(II)

17. a. b. c.

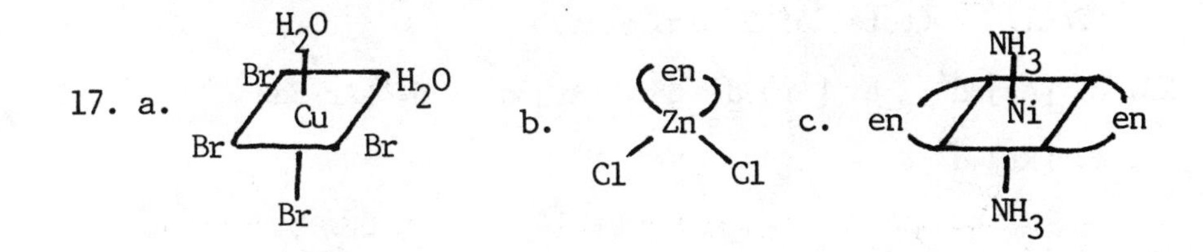

d. ox Cu ox ox

e. H_2O Cl Ni Cl H_2O

19. acac acac Fe acac

21. a. NH_3 SCN SCN Cr SCN SCN NH_3 — NH_3 SCN NH_3 Cr SCN SCN SCN

b. NH_3 NO_2 NH_3 Co NO_2 NH_3 NO_2 — NH_3 NO_2 NH_3 Co NO_2 NO_2 NH_3

c. 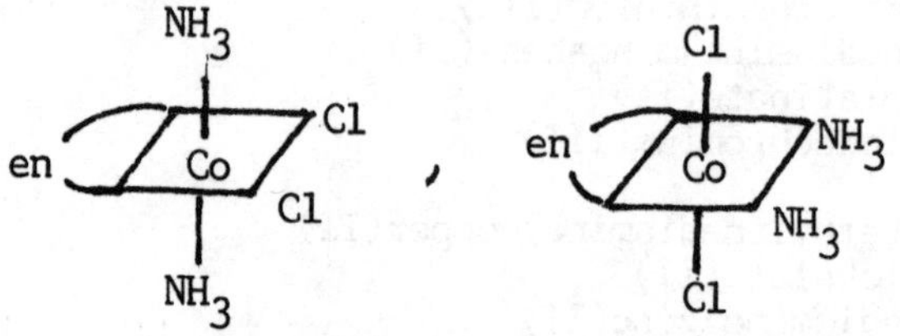 NH3, Cl, Cl, NH3, en, Co ; Cl, NH3, NH3, Cl, en, Co

23. $Fe(C_2O_4)_3^{3-}$; $Fe(C_2O_4)_2Cl_2^{3-}$ (2 isomers); $Fe(C_2O_4)Cl_4^{3-}$; $FeCl_6^{3-}$ (total of 5 complexes)

25. a. $[Ar]3d^1$ b. $[Ar]3d^4$ c. $[Ar]3d^7$ d. $[Ar]3d^7$
e. $[Kr]4d^4$

		3d	
27.	a. [Ar]	(↑)( )( )( )( )	1
	b. [Ar]	(↑)(↑)(↑)(↑)( )	4
	c. [Ar]	(↑↓)(↑↓)(↑)(↑)(↑)	3
	d. [Ar]	(↑↓)(↑↓)(↑)(↑)(↑)	3
		4d	
	e. [Kr]	(↑)(↑)(↑)(↑)( )	4

29. a. Co^{3+}: $3d^6$ b. Mn^{2+}: $3d^5$

a.		b.	
( )( )	(↑)(↑)	( )( )	(↑)(↑)
(↑↓)(↑↓)(↑↓)	(↑↓)(↑)(↑)	(↑↓)(↑↓)(↑)	(↑)(↑)(↑)
low spin	high spin	low spin	high spin

31. Mn^{3+} is $3d^4$; there are two different ways to distribute 4 e^-
Mn^{4+} is $3d^3$; there is only one way to distribute 3 e^-

33. In $Cr(CN)_6^{4-}$, Cr^{2+} ($3d^4$) is low spin: ( )( ) / (↑↓)(↑)(↑)
In $Cr(H_2O)_6^{2+}$, Cr^{2+} is high spin: (↑)( ) / (↑)(↑)(↑)

35. low-spin complexes
a. 2 b. 0 c. 0 d. 2 e. 2

37. $$\Delta_o = \frac{(6.626 \times 10^{-34}\ J\cdot s)(2.998 \times 10^{8}\ m/s)}{460 \times 10^{-9}\ m} \times \frac{1\ kJ}{10^3\ J}$$
$$\times\ 6.022 \times 10^{23}/mol = 2.60 \times 10^{2}\ kJ/mol$$

39. a. Fe   b. Mg   c. Co

41. $K = 200 = \frac{[\text{Hem CO}] \times P\,O_2}{[\text{Hem } O_2] \times P\,CO} = \frac{0.050}{0.950} \times \frac{P\,O_2}{P\,CO}$

$P\,CO/P\,O_2 = 2.6 \times 10^{-4}$

43. a. two unshared pairs of electrons on two nitrogens
b. forms $Ag(NH_3)_2^{+}$
c. all positions equivalent

45. a. $[Pt(NH_3)_4Cl_2]Cl_2$; true   b. true   c. true

47. a. n Co = 22.0/58.93 = 0.373
n N = 31.4/14.01 = 2.24
n H = 6.78/1.008 = 6.73 $CoN_6H_{18}Cl_3$
n Cl = 39.8/35.45 = 1.12

b. $[Co(NH_3)_6]Cl_3(s) \longrightarrow Co(NH_3)_6^{3+}(aq) + 3Cl^-(aq)$

49. $Pt(NH_3)_4^{2+}$ cation, $PtCl_4^{2-}$ anion <u>or</u>

$Pt(NH_3)_3Cl^+$ cation, $Pt(NH_3)Cl_3^-$ anion

50. a. n Cu = 20.25/63.55 = 0.3186
n C = 15.29/12.01 = 1.273
n H = 7.07/1.008 = 7.01 $CuC_4H_{22}N_6SO_4$; en = $C_2N_2H_8$
n N = 26.86/14.01 = 1.917
n S = 10.23/32.07 = 0.3190
n O = 20.39/16.00 = 1.274

b. $Cu(en)_2(NH_3)_2$; cis and trans

51. a. $Al(OH)_3(s) + OH^-(aq) \longrightarrow Al(OH)_4^-(aq)$
b. $Al^{3+}(aq) + 4\,OH^-(aq) \longrightarrow Al(OH)_4^-(aq)$
c. $K = K_b \times K_{sp} = (1 \times 10^{33})(2 \times 10^{-31}) = 2 \times 10^2$

52. $Cu^{2+}(aq) + 2NH_3(aq) + 2H_2O \longrightarrow Cu(OH)_2(s) + 2NH_4^+(aq)$

$Cu(OH)_2(s) + 4NH_3(aq) \longrightarrow Cu(NH_3)_4^{2+}(aq) + 2\,OH^-(aq)$

$Cu(NH_3)_4^{2+}(aq) + 4H^+(aq) + 4Cl^-(aq) \longrightarrow CuCl_4^{2-}(aq) + 4NH_4^+(aq)$

CHAPTER 16
Spontaneity of Reaction

LECTURE NOTES

This chapter takes an empirical approach to the thermodynamic functions H, S, and G. Emphasis is placed on the effect of $\Delta H°$, $\Delta S°$, and $\Delta G°$ on reaction spontaneity. We do not attempt to derive the Gibbs-Helmholtz equation, nor do we discuss how molar entropy is determined. These aspects of thermodynamics can safely be deferred to a later course. It is important that students, in their first exposure, appreciate the power of thermodynamics, even at the expense of its rigor or elegance.

Perhaps surprisingly, students find this material quite straightforward. It is readily covered in two lectures.

<u>LECTURE 1</u>

I <u>Spontaneous Reactions</u>

A. Examples:

$CH_4(g) + 2\ O_2(g) \longrightarrow CO_2(g) + 2H_2O(l)$

$H_2O(s) \longrightarrow H_2O(l)$ (at 25°C)

B. Factors affecting
1. Energy factor: at 25°C, 1 atm, exothermic reactions are ordinarily spontaneous ($\Delta H < 0$).
2. Randomness factor: other things being equal, system tends to move from a more ordered to a more random structure.

II <u>Entropy Changes</u>

A. $\Delta S = S_{products} - S_{reactants}$; measure of change in order

solid $\longrightarrow$ liquid; ΔS positive

liquid $\longrightarrow$ gas; ΔS positive

ΔS is usually positive for a reaction in which the number of moles of gas increases.

$2SO_3(g) \longrightarrow 2SO_2(g) + O_2(g)$; $\Delta n_g = +1$; ΔS positive

$N_2(g) + 3H_2(g) \longrightarrow 2NH_3(g)$ $\Delta n_g = -2$; ΔS negative

B. Calculation of $\Delta S°$ (ΔS at 1 atm, 1 M)

$$\Delta S° = \sum S°_{products} - \sum S°_{reactants}$$

$Fe_2O_3(s) + 3H_2(g) \longrightarrow 2Fe(s) + 3H_2O(g)$

$$\Delta S° = 2S° \ Fe(s) + 3S° \ H_2O(g) - S° \ Fe_2O_3(s) - 3S° \ H_2(g)$$

$$= 2(27.2 \ J/K) + 3(188.7 \ J/K) - 90.0 \ J/K - 3(130.6 \ J/K)$$

$$= 138.7 \ J/K$$

Note that $S°$ is a positive quantity for both elements and compounds.

C. Reactions for which $\Delta S°$ is positive tend to be spontaneous, at least at high temperatures.

$H_2O(s) \longrightarrow H_2O(l)$ $\Delta S° > 0$

$H_2O(l) \longrightarrow H_2O(g)$ $\Delta S° > 0$

$Fe_2O_3(s) + 3H_2(g) \longrightarrow 2Fe(s) + 3H_2O(g)$ $\Delta S° > 0$

All of these reactions are endothermic ($\Delta H > 0$). They become spontaneous at high temperatures.

III Free Energy Changes

A. $\Delta G° = \Delta H° - T\Delta S°$

Note that ΔG, like ΔS, is dependent on pressure, concentration. Unlike $\Delta H°$ and $\Delta S°$, $\Delta G°$ is strongly temperature dependent because of T in equation.

If $\Delta G° < 0$, reaction spontaneous at standard conditions

If $\Delta G° > 0$, reaction nonspontaneous at standard cond.

If $\Delta G° = 0$, reaction at equilibrium at standard cond.

B. Effect of $\Delta H°$, $\Delta S°$ on spontaneity

If $\Delta H° > 0$, $\Delta S° < 0$, $\Delta G° > 0$ at all T, nonspontaneous

If $\Delta H° < 0$, $\Delta S° > 0$, $\Delta G° < 0$ at all T, spontaneous

If $\Delta H° > 0$, $\Delta S° > 0$, $\Delta G° > 0$ at low T, < 0 at high T
spontaneous at high T

If $\Delta H° < 0$, $\Delta S° < 0$, $\Delta G° < 0$ at low T, > 0 at high T
nonspontaneous at high T

LECTURE 2

I Free Energy Change

A. Calculation of $\Delta G°$ from $\Delta H°$ and $\Delta S°$

1. $Fe_2O_3(s) + 3H_2(g) \longrightarrow 2Fe(s) + 3H_2O(g)$

   $\Delta H° = +96.8$ kJ; $\Delta S° = +138.7$ J/K $= +0.1387$ kJ/K

   $\Delta G°$ (in kJ) $= +96.8 - 0.1387\ T$

   at 25°C: $\Delta G° = +96.8 - 0.1387(298) = +55.5$ kJ
   at 500°C: $\Delta G° = +96.8 - 0.1387(773) = -10.4$ kJ
   nonspontaneous at 25°C, spontaneous at 500°C

2. At what temperature does the redction of Fe_2O_3 by H_2 become spontaneous at 1 atm?

   $\Delta G° = 0$; $\Delta H° = T\Delta S°$

   $T = \Delta H°/\Delta S° = 96.8/0.1387 = 698$ K (425°C)

B. Calculation of $\Delta G°$ at 25°C from $\Delta G_f°$

$\Delta G° = \sum \Delta G_f°$ products $- \sum \Delta G_f°$ reactants

$PbCl_2(s) \longrightarrow Pb^{2+}(aq) + 2Cl^-(aq)$

$\Delta G° = \Delta G_f°\ Pb^{2+} + 2\,\Delta G_f°\ Cl^- - \Delta G_f°\ PbCl_2 = +27.3$ kJ

C. Calculation of ΔG from $\Delta G°$

$\Delta G = \Delta G° + RT \ln Q$

$PbCl_2(s) \longrightarrow Pb^{2+}(0.0010\ M) + 2Cl^-(0.001\ M)$

$\Delta G = \Delta G° + (0.00831)(298) \ln (0.0010)(0.0010)^2$
$= +27.3$ kJ $- 51.3$ kJ $= -24.0$ kJ

Note that reaction is spontaneous at this low concentration

D. Relation between $\Delta G°$ and K

$\Delta G° = -RT \ln K$

If $K > 1$, $\Delta G° < 0$, reaction spontaneous at stand. cond.
If $K < 1$, $\Delta G° > 0$, reaction nonspontaneous at stand. cond.
If $K = 1$, $\Delta G° = 0$, reaction at equilibrium at stand. cond.

$H_2O(l) \longrightarrow H^+(aq) + OH^-(aq)$

$\Delta G° = -(0.00831)(298) \ln (1.0 \times 10^{-14}) = +79.8$ kJ

Nonspontaneous at standard conditions, (1 M H^+, 1 M OH^-)

DEMONSTRATIONS

1. Spontaneous endothermic reactions: J. Chem. Educ. 46 A55 (1969), 51 A178 (1974)
2. Maximum work: J. Chem. Educ. 67 962 (1990)
3. Reaction of copper(II) oxide with hydrogen: Shakhashiri Videotapes, Demonstration 9.

QUIZZES

Quiz 1

1. Consider the reaction:

   $SnO_2(s) + 2H_2(g) \longrightarrow Sn(s) + 2H_2O(g)$

   calculate

   a. $\Delta H°$ b. $\Delta S°$ c. $\Delta G°$ at 25°C d. $\Delta G°$ at 200°C

2. For which of the following processes would you expect ΔS to be positive?

   a. osmosis b. $NO(g) + \frac{1}{2} O_2(g) \longrightarrow NO_2(g)$

   c. compressing a gas

Quiz 2

1. For the reaction: $N_2(g) + 3H_2(g) \longrightarrow 2NH_3(g)$, calculate

   a. $\Delta H°$ b. $\Delta S°$ c. $\Delta G°$ at 200°C

   b. the temperature at which $\Delta G° = 0$

2. Taking K_{sp} of $PbCl_2$ to be 1.7×10^{-5}, calculate $\Delta G°$ at 25°C for the reaction:

   $Pb^{2+}(aq) + 2Cl^-(aq) \longrightarrow PbCl_2(s)$

Quiz 3

1. Consider the reaction: $CaSO_4(s) \longrightarrow Ca^{2+}(aq) + SO_4^{2-}(aq)$

   a. Calculate $\Delta G°$ at 25°C from free energies of formation.

   b. Calculate $\Delta G°$ at 25 C from the Gibbs-Helmholtz equation.

   c. Calculate ΔG at 25°C when $[Ca^{2+}] = [SO_4^{2-}] = 0.010$ M

2. Which of the following are spontaneous at 25°C, 1 atm?
   a. dissolving HCl in water to form a 1 M solution.
   b. rusting of iron.
   c. decomposition of water to the elements.

Quiz 4

1. Consider the reaction between Fe_3O_4 and hydrogen gas to form iron and water vapor. Write a balanced equation for the reaction and calculate $\Delta G°$ at 300°C.

2. What are the signs of ΔS and ΔH for a reaction that:
   a. is spontaneous at all temperatures?
   b. becomes spontaneous as temperature increases?

Quiz 5

1. Consider the reaction: $N_2(g) + 3H_2(g) \rightarrow 2NH_3(g)$
   a. Calculate $\Delta H°$ and $\Delta S°$ for this reaction.
   b. Calculate $\Delta G°$ at 200°C.
   c. Calculate ΔG for this reaction at 300°C when all species are at a partial pressure of 0.010 atm.

2. Calculate K at 100°C for the reaction: $H_2O(l) \rightleftharpoons H_2O(g)$
   Take $\Delta H° = +44.3$ kJ, $\Delta S° = +0.1188$ kJ/K

Answers

Quiz 1 1. a. +97.1 kJ b. +115.5 J/K c. +62.7 kJ d. +42.5 kJ
2. a

Quiz 2 1. a. -92.2 kJ b. -198.7 J/K c. +1.8 kJ d. 464 K
2. -27.2 kJ

Quiz 3 1. a. +23.7 kJ b. +23.7 kJ c. +0.9 kJ
2. a, b

Quiz 4 1. $Fe_3O_4(s) + 4H_2(g) \rightarrow 3Fe(s) + 4H_2O(g)$; 55.0 kJ
2. a. ΔH - , ΔS + b. ΔH + , ΔS +

Quiz 5 1. a. $\Delta H° = -92.2$ kJ $S° = -198.7$ J/K b. +1.8 kJ
c. 21.7 kJ + 43.9 kJ = +65.6 kJ

2. 1.0

PROBLEMS

1. a, c

3. a, c spontaneous

5. a. - b. + c. - d. -

7. a. - b. + c. - d. -

9. a. - b. + c. - d. -

11.a. $\Delta S° = 4S°\ NO_2(g) + 6S°\ H_2O(g) - 7S°\ O_2(g) - 4S°\ NH_3(g)$
$= -112.0$ J/K

b. $\Delta S° = S°\ N_2(g) + 4S°\ H_2O(g) - S°\ N_2H_4(l) - 2S°\ H_2O_2(l)$
$= +605.9$ J/K

c. $\Delta S° = S°\ CO_2(g) - S°\ C(s) - S°\ O_2(g) = +2.9$ J/K

d. $\Delta S° = S°\ CHCl_3(l) + 3S°\ HCl(g) - S°\ CH_4(g) - 3S°\ Cl_2(g)$
$= -93.1$ J/K

13. a. $\Delta S° = S°\ Zn^{2+}(aq) + S°\ H_2(g) - S°\ Zn(s) = -23.1$ J/K

b. $\Delta S° = S°\ H_2O(l) - S°\ OH^-(aq) = +80.7$ J/K

c. $\Delta S° = S°\ NH_4^+(aq) + S°\ OH^-(aq) - S°\ NH_3(g) - S°\ H_2O(l)$
$= -159.6$ J/K

15. a. $\Delta S° = 2S°\ H_2O(g) + 2S°\ SO_2(g) - 2S°\ H_2S(g) - 3S°\ O_2(g)$
$= -152.8$ J/K

b. $\Delta S° = S°\ Ag^+(aq) + S°\ H_2O(l) + S°\ NO_2(g) - S°\ Ag(s)$
$- S°\ NO_3^-(aq) = +193.6$ J/K

17. a. $\Delta G° = +210$ kJ - (298 K)(0.0325 kJ/K) = +200 kJ

b. $\Delta G° = +638$ kJ + (298 K)(0.2152 kJ/K) = +702 kJ

c. $\Delta G° = +7.34$ kJ - (298 K)(0.337 kJ/K) = -93 kJ

19. a. $\Delta H° = 4(+33.2$ kJ) + 6(-241.8 kJ) - 4(-46.1 kJ) = -1133.6 kJ

$\Delta G° = -1133.6$ kJ - 673 K(-0.1120 kJ/K) = -1058.2 kJ; spont.

b. $\Delta H° = 4(-241.8$ kJ) - 50.6 kJ + 2(187.8 kJ) = -642.2 kJ

$\Delta G° = -642.2$ kJ - 673 K(+0.6059 kJ/K) = -1050.0 kJ; spont.

c. $\Delta H° = -393.5$ kJ

$\Delta G° = -393.5$ kJ - 673 K(+0.0029 kJ/K) = -395.5 kJ; spont.

d. $\Delta H° = 3(-92.3$ kJ) - 134.5 kJ + 74.8 kJ = -336.6 kJ

$\Delta G° = -336.6\ kJ - 673\ K(-0.0931\ kJ/K) = -273.9\ kJ$; spont.

21. a. $\Delta G° = 4(+51.3\ kJ) + 6(-228.6\ kJ) - 4(-16.5\ kJ) = -1100.4\ kJ$

    b. $\Delta G° = 4(-228.6\ kJ) - 149.2\ kJ - 2(-120.4\ kJ) = -822.8\ kJ$

    c. $\Delta G° = -394.4\ kJ$

    d. $\Delta G° = 3(-95.3\ kJ) - 73.7\ kJ + 50.7\ kJ = -308.9\ kJ$

23. a. $Ca(s) + C(s) + 3/2\ O_2(g) \longrightarrow CaCO_3(s)$

    $\Delta H° = -1206.9\ kJ \qquad \Delta S° = -0.2617\ kJ/K$

    $\Delta G_f° = -1206.9\ kJ + 298(0.2617\ kJ) = -1128.9\ kJ$; spont.

    b. $Mn(s) + \frac{1}{2}\ O_2(g) \longrightarrow MnO(s)$

    $\Delta H° = -385.2\ kJ \qquad \Delta S° = -0.0748\ kJ/K$

    $\Delta G_f° = -385.2\ kJ + 298(0.0748\ kJ) = -362.9\ kJ$; spont.

    c. $1/4\ P_4(s) + 5/2\ Cl_2(g) \longrightarrow PCl_5(g)$

    $\Delta H° = -374.9\ kJ \qquad \Delta S° = -0.2341\ kJ/K$

    $\Delta G_f° = -374.9\ kJ + 298(0.2341\ kJ) = -305.1\ kJ$; spont.

    all are stable

25. $\Delta G° = -134.0\ kJ + 237.2\ kJ - 163.2\ kJ = -60.0\ kJ$; plausible

27. a. $\Delta S° = \dfrac{\Delta H° - \Delta G°}{T} = \dfrac{+101.0\ kJ}{298\ K} = +0.339\ kJ/K$; yes, $\Delta n_g = +2$

    b. 0 K; +135.6 kJ
    1000 K; +135.6 kJ - 339 kJ = -203 kJ

29. a. $\Delta G° = -54.3\ kJ - 298\ K(+0.1251\ kJ/K) = -91.6\ kJ$

    b. $-54.3\ kJ = -168.6\ kJ - 334.4\ kJ - 285.8\ kJ + 460.0\ kJ$
    $-2\ mol\ \Delta H_f°\ CuCl$

    $\Delta H_f°\ CuCl = -137.2\ kJ/mol$

    c. $+125.1\ J/K = +93.1\ J/K + 113.0\ J/K + 69.9\ J/K + 21.6\ J/K$
    $- 2\ mol\ (S°\ CuCl)$

    $S°\ CuCl = +86.2\ J/mol \cdot K$

31. a. becomes more nearly spontaneous as T rises
    b. spontaneous at all T
    c. becomes spontaneous at high T

33. a. 851.5/0.0385 = 2.22 x 10^4 K; impossibly high T

b. -50.6/0.3315 = -153 K; no temperature

c. 98.9/0.0939 = 1050 K; becomes spontaneous at 1050 K

35. $\Delta H°$ = +176.0 kJ; $\Delta S°$ = +0.2845 kJ

T = 176.0 K/0.2845 = 619 K

37. a. $\Delta H°$ = +62.0 kJ $\Delta S°$ = +0.1328 kJ/K

$\Delta G°$ = +62.0 kJ - 0.1328 kJ/K x T

T(K)	100	200	300	400	500
$\Delta G°$(kJ)	+48.7	+35.4	+22.2	+8.9	-4.4

b. 62.0/0.1328 = 467 K

39. a. $\Delta H°$ = +234.0 kJ $\Delta S°$ = +0.2790 kJ/K

T = 234.0/0.2790 = 839 K

b. $\Delta H°$ = +98.8 kJ $\Delta S°$ = +0.1415 kJ/K

T = 98.8/0.1415 = 698 K

(b) is lower

41. P(red) $\longrightarrow$ P(white)

$\Delta H°$ = +17.6 kJ $\Delta S°$ = +18.29 J/K T = 962 K

43. Hg(l) $\longrightarrow$ Hg(g)

$\Delta H°$ = +61.32 kJ $\Delta S°$ = +99.0 J/K T = 619 K = 346°C

45. a. ΔG = $\Delta G°$ = +18.0 kJ; no

b. ΔG = +18.0 kJ + RT ln(1.0 x 10^{-6}) = -16.2 kJ; yes

47. a. $\Delta G°$ = -314.4 kJ + 262.4 kJ + 474.4 kJ = +422.4 kJ

b. $\Delta G = +422.2 \text{ kJ} + RT \ln \dfrac{(1.8 \times 10^{-2})^2 \times 0.200 \times 0.200}{(0.500)^2}$

= -398.0 kJ

49. a. $\Delta G°$ = +77.1 kJ - 131.2 kJ + 109.8 kJ = +55.7 kJ

b. -1.0 = +55.7 + (0.00831)(298) ln x^2

x = $[Ag^+]$ = $[Cl^-]$ = 1.1 x 10^{-5} M

c. yes; $x^2 \approx K_{sp}$

51. $2FeCl_2(s) \longrightarrow 2Fe(s) + 2Cl_2(g)$ $\Delta G° = +604.6$ kJ

$2Fe(s) + 3Cl_2(g) \longrightarrow 2FeCl_3(s)$ $\Delta G° = -668.0$ kJ

$2FeCl_2(s) + Cl_2(g) \longrightarrow 2FeCl_3(s)$ $\Delta G° = -63.4$ kJ

53. 372/31 = 12

55. a. $\Delta G° = -157.2 \text{ kJ} + 237.2 \text{ kJ} = +80.0 \text{ kJ}$

b. $\ln K_w = -80.0/(0.00831 \times 298) = -32.3$; $K_w = 1 \times 10^{-14}$

57. a. $\Delta G° = -(0.00831)(298) \ln(1.1 \times 10^{-16}) = +91.0 \text{ kJ}$

b. $\Delta G° = +70.4$ kJ c. $\Delta G° = -15.9$ kJ

59. $HF(aq) \rightleftharpoons H^+(aq) + F^-(aq)$

$\Delta H° = -12.5$ kJ $\Delta S° = -102.5$ J/K $\Delta G° = +18.0$ kJ

$\ln K_a = -18.0/(0.00831 \times 298) = -7.27$; $K_a = 7.0 \times 10^{-4}$

61. See p. 453

63. less efficient

65. a. independent of T, P b. independent of T c. neither
d. neither

67. a. less gas, so less random
b. $\Delta S° = \sum S°_{products} - \sum S°_{reactants}$
c. solid more ordered

69. $K = \dfrac{2.65 \times 1.28}{1.84 \times 2.81} = 0.656$

$\Delta G° = -(0.00831)(1000) \ln 0.656 = +3.50 \text{ kJ}$

71. $PbCl_2(s) \rightleftharpoons Pb^{2+}(aq) + 2Cl^-(aq)$

$\Delta H° = +23.3$ kJ $\Delta S° = -12.5$ J/K $\Delta G° = +28.0$ kJ

$\ln K_{sp} = -28.0/(0.00831 \times 373)$ $K_{sp} = 1.2 \times 10^{-4}$

$4s^3 = 1.2 \times 10^{-4}$ $s = 0.031$ mol/L

$m = \dfrac{0.031 \text{ mol}}{1 \text{ L}} \times \dfrac{278.1 \text{ g}}{1 \text{ mol}} \times 0.200 \text{ L} = 1.7 \text{ g}$

73. $C_6H_6(l) \longrightarrow C_6H_6(g)$

$\Delta S° = +95.4$ J/K $\Delta G° = 0$ at 353 K $\Delta H° = T\Delta S° = +33.7$ kJ

$\ln 760/150 = 33{,}700[1/T_1 - 1/353]/8.31$

$T_1 = 309\ K = 36°C$

74. $\Delta H° = +62.4\ kJ - 53.0\ kJ = +9.4\ kJ$

$\Delta S° = 260.7\ J/K + 130.6\ J/K - 413.0\ J/K = -21.7\ J/K$

$\Delta G° = +9.4\ kJ + 793\ K(0.0217\ kJ/K) = +26.6\ kJ$

$\ln K = -26.6/(0.00831 \times 793) = -4.04;\ K = 0.018$

$\frac{(0.100 - x)^2}{(0.100 + 2x)^2} = 0.018 \quad \frac{0.100 - x}{0.100 + 2x} = 0.13 \qquad x = 0.069$

$P\ H_2 = P\ I_2 = 0.031\ atm; \quad P\ HI = 0.238\ atm$

75. a. $\Delta H° = 333\ J/g \times 18.02\ g/mol = 6.00 \times 10^3\ J = 6.00\ kJ$

b. 0

c. $\Delta S° = \Delta H°/T = 6.00\ kJ/273\ K = +0.0220\ kJ/K$

d. $6.00\ kJ - 263\ K(0.0220\ kJ/K) = +0.21\ kJ$

e. $6.00\ kJ - 283\ K(0.0220\ kJ/K) = -0.23\ kJ$

76. a. $\Delta S° = \frac{\Delta H° - \Delta G°}{T} = +140\ kJ/298\ K = +0.470\ kJ/K$

$\Delta G°$ at 37°C $= -5650\ kJ - 310\ K(0.470\ kJ/K) = -5800\ kJ$

$1\ g \times \frac{1\ mol}{342.30\ g} \times 5800\ \frac{kJ}{mol} \times 0.30 = 5.1\ kJ$

b. Assume 130 lb $\approx$ 60 kg

$w = 9.79 \times 10^{-3} \times 60 \times 1610 = 9.5 \times 10^2\ kJ$

$9.5 \times 10^2/5.1 \approx 1.9 \times 10^2\ g$

77. $CaH_2(s) \rightarrow Ca(s) + H_2(g)$

$\Delta H° = +186.2\ kJ$

$\Delta S° = +0.0414\ kJ/K + 0.1306\ kJ/K - 0.0420\ kJ/K = +0.1300\ kJ/K$

$T = 186.2/0.1300 = 1430\ K \approx 1160\ °C$

78. $2CuO(s) \rightarrow Cu_2O(s) + \frac{1}{2}\ O_2(g)$

$\Delta H° = +146.0\ kJ \quad \Delta S° = +0.1104\ kJ/K \quad T = 1322\ K \approx 1050°C$

$Cu_2O(s) \rightarrow 2Cu(s) + \frac{1}{2}\ O_2(g)$

$\Delta H° = +168.6\ kJ \quad \Delta S° = +0.0758\ kJ/K \quad T = 2220\ K \approx 1950°C$

CHAPTER 17
Electrochemistry

LECTURE NOTES

The most difficult portion of this chapter involves the Nernst equation. Students have a great deal of trouble using this equation to calculate the concentration of a species (e.g., H^+), knowing E and E°. The principal problem here is their unfamiliarity with logarithms. Other topics in this chapter, such as applications of E°, ΔG° and K, electrolysis calculations, etc. seem to go smoothly.

You should expect to spend at least 2½ lectures on this chapter, 3 if you want to discuss commercial cells in any detail.

LECTURE 1

I Voltaic Cells

Spontaneous reaction used to produce electrical energy.

A. Salt bridge cells

$Zn(s) + Cu^{2+}(aq) \longrightarrow Zn^{2+}(aq) + Cu(s)$

must design cell to make electron transfer occur indirectly

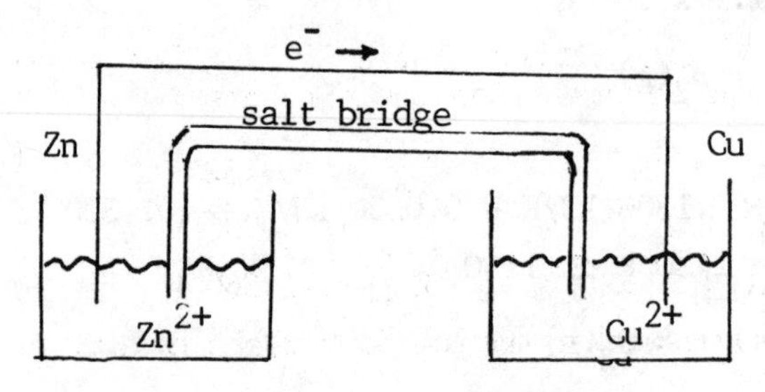

anode: $Zn(s) \longrightarrow Zn^{2+}(aq) + 2\ e^-$

cathode: $Cu^{2+}(aq) + 2\ e^- \longrightarrow Cu(s)$

Salt bridge allows current to flow but prevents contact between Zn and Cu^{2+}, which would shortcircuit the cell.

$Zn(s) + 2H^+(aq) \longrightarrow Zn^{2+}(aq) + H_2(g)$

use inert, Pt electrode for H^+, H_2 half-cell

B. Cell Notation

$Zn - Cu^{2+}$ cell: $Zn \mid Zn^{2+} \parallel Cu^{2+} \mid Cu$

$Zn - H^+$ cell $Zn \mid Zn^{2+} \parallel H^+ \mid H_2 \mid Pt$

II Standard Voltages

A. $E°$ = cell voltage when all species are at standard concentrations (1 atm for gases, 1 M for solutes in water)

$Zn(s) + 2H^+(aq, 1 M) \longrightarrow Zn^{2+}(aq, 1 M) + H_2 (g, 1 atm)$

$E° = +0.762 V = E°_{ox} Zn + E°_{red} H^+$

arbitrarily, set $E°_{red} H^+ = 0.000 V$; thus $E°_{ox} Zn = +0.762 V$

Table 17.1 lists values of $E°_{red}$; can get $E°_{ox}$ by changing sign

$Cu(s) \longrightarrow Cu^{2+}(aq) + 2e^-$ $E°_{ox} Cu = -E°_{red} Cu^{2+} = -0.339 V$

$Zn(s) \longrightarrow Zn^{2+}(aq) + 2e^-$ $E°_{ox} Zn = -E°_{red} Zn^{2+} = +0.762 V$

B. Relative strengths of oxidizing and reducing agents

1. Oxidizing agents (left column, Table 17.1). The larger (more positive) the value of $E°_{red}$, the stronger the oxidizing agent.

$Zn^{2+}(aq) + 2e^- \longrightarrow Zn(s)$ $E°_{red} = -0.762 V$

$2H^+(aq) + 2e^- \longrightarrow H_2(g)$ $E°_{red} = 0.000 V$

$Cl_2(g) + 2e^- \longrightarrow 2Cl^-(aq)$ $E°_{red} = +1.360 V$

oxidizing agents become stronger moving down left column

2. Reducing agents (right column, Table 17.1). The larger the value of $E°_{ox}$, the stronger the reducing agent.

$Zn(s) \longrightarrow Zn^{2+}(aq) + 2e^-$ $E°_{ox} = +0.762 V$

$H_2(g) \longrightarrow 2H^+(aq) + 2e^-$ $E°_{ox} = 0.000 V$

$2Cl^-(aq) \longrightarrow Cl_2(g) + 2e^-$ $E°_{ox} = -1.360 V$

reducing agents become weaker moving down right column

C. Calculation of $E°$

$E° = E°_{ox} + E°_{red}$

$Cl_2(g) + 2Br^-(aq) \longrightarrow 2Cl^-(aq) + Br_2(l)$

$E° = E°_{red} Cl_2 + E°_{ox} Br^- = 1.360 V - 1.077 V = +0.283 V$

Since calculated voltage is positive, this reaction can occur in a voltaic cell.

D. Determination of whether redox reaction will occur

What, if anything, happens when bromine is added to a solution of tin(II) chloride?

Possible oxidations:

$Sn^{2+}(aq) \longrightarrow Sn^{4+}(aq) + 2e^-$ $\qquad E^\circ_{ox} = -0.154$ V

$2Cl^-(aq) \longrightarrow Cl_2(g) + 2e^-$ $\qquad E^\circ_{ox} = -1.360$ V

Possible reductions:

$Sn^{2+}(aq) + 2e^- \longrightarrow Sn(s)$ $\qquad E^\circ_{red} = -0.141$ V

$Br_2(l) + 2e^- \longrightarrow 2Br^-(aq)$ $\qquad E^\circ_{red} = +1.077$ V

Reaction:

$Sn^{2+}(aq) + Br_2(l) \longrightarrow Sn^{4+}(aq) + 2Br^-(aq)$; $E^\circ = +0.923$

LECTURE 2

I Relation Between E°, ΔG° and K

$\Delta G^\circ = -nFE^\circ \qquad \ln K = nE^\circ/0.0257$

Note that if E° is +, ΔG° is -, ln K is +, K > 1

$Cl_2(g) + 2Br^-(aq) \longrightarrow 2Cl^-(aq) + Br_2(l)$

$E^\circ = 1.360 \text{ V} - 1.077 \text{ V} = +0.283 \text{ V}$

$\Delta G^\circ = -2(96.5)(+0.283)\text{kJ} = -54.6 \text{ kJ}$

$\ln K = \frac{2(0.283)}{0.0257} = 22.0; \; K = 4 \times 10^9$

II Effect of Concentration upon Voltage

A. Nernst Equation

$E = E^\circ + \frac{RT}{nF} \ln Q \quad = E^\circ - \frac{0.0257}{n} \ln Q$

In expression for Q, gases enter as partial pressures in atmospheres, solutes as concentrations in moles per liter

$Cl_2(g) + 2Br^-(aq) \longrightarrow 2Cl^-(aq) + Br_2(l)$

$E = +0.283 \text{ V} - \frac{0.0257}{2} \ln \frac{[Cl^-]^2}{(P\, Cl_2) \times [Br^-]^2}$

Calculate voltage when $[Br^-]$ = 1 M, P Cl_2 = 1 atm, $[Cl^-]$ = 0.01 M.

$$E = +0.283\ V - \frac{0.0257}{2} \ln (0.01)^2 = +0.401\ V$$

B. Use of Nernst equation to determine concentrations of ions in solution.

$$Zn(s) + 2H^+(aq) \longrightarrow Zn^{2+}(aq) + H_2(g)$$

$$E = +0.762\ V - \frac{0.0257}{2} \ln \frac{(P\ H_2) \times [Zn^{2+}]}{[H^+]^2}$$

Suppose $[Zn^{2+}]$ = 1 M, P H_2 = 1 atm

$$E = +0.762\ V + 0.0257 \ln [H^+]$$

Measure voltage, calculate $[H^+]$. Suppose E = 0.200 V

$$\ln [H^+] = \frac{-0.562}{0.0257} = -21.9;\ [H^+] = 3 \times 10^{-10}\ M$$

LECTURE 2½

I Electrolytic Cells

Electrical energy supplied to bring about a nonspontaneous redox reaction. Cell diagram.

A. Amount of products formed in electrolysis

1. $Ag^+(aq) + e^- \longrightarrow Ag(s)$ 1 mol e^- = 96480 C $\longrightarrow$ 1 mol Ag

   no. of coulombs = no. of amperes x no. of seconds

   no. of joules = no. of coulombs x no. of volts

2. How much silver is plated from $AgNO_3$ solution by a current of 2.60 A in one hour?

$$n\ e^- = (2.60)(3600)C \times \frac{1\ mol\ e^-}{96480\ C} = 0.0970\ mol\ e^-$$

$$m\ Ag = 0.0970\ mol\ e^- \times \frac{1\ mol\ Ag}{1\ mol\ e^-} \times \frac{107.9\ g\ Ag}{1\ mol\ Ag} = 10.5\ g\ Ag$$

II Commercial Cells

A. Electrolysis of aqueous NaCl

$$2H_2O + 2Cl^-(aq) \longrightarrow H_2(g) + Cl_2(g) + 2\ OH^-(aq)$$

Voltage required = 1.360 V + 0.828 V = 2.188 V

Energy required to form one mole of Cl_2?

$$\text{no. coulombs} = 1 \text{ mol } Cl_2 \times \frac{2 \text{ mol } e^-}{1 \text{ mol } Cl_2} \times \frac{96480 \text{ C}}{1 \text{ mol } e^-}$$

$$= 1.930 \times 10^5 \text{ C}$$

$$\text{no. joules} = (1.930 \times 10^5)(2.188) = 4.223 \times 10^5 \text{ J} = 422.3 \text{ kJ}$$

(0.1173 kWh)

B. Lead Storage Battery

anode: $Pb(s) + SO_4^{2-}(aq) \longrightarrow PbSO_4(s) + 2e^-$

cathode: $PbO_2(s) + 4H^+(aq) + SO_4^{2-}(aq) + 2e^- \longrightarrow PbSO_4(s) + 2H_2O$

overall: $Pb(s) + PbO_2(s) + 4H^+(aq) + 2SO_4^{2-}(aq) \longrightarrow 2PbSO_4(s) + 2H_2O$

As cell discharges, conc. H_2SO_4 and density decrease

DEMONSTRATIONS

1. Human salt bridge: J. Chem. Educ. 67 156 (1990)
2. Vegetable and fruit juice voltages: J. Chem. Educ. 65 727 (1988)
3. The voltaic pile: J. Chem. Educ. 68 665 (1991)
4. Half-cell reactions: J. Chem. Educ. 68 247 (1991)
5. Metal trees: Test. Dem. 53, 127
6. Zn - H^+ cell: Test. Dem. 141; Shak. 4 130
7. Measurement of potentials: Test. Dem. 196
8. Electrolytic and voltaic cells: Test. Dem. 20, 81, 150
9. Concentration cells: Test. Dem. 144; Shak. 4 140
10. Electrolysis of water solutions: Test. Dem. 78, 129, 160, 161; Shak. 4 170-181, 205-209
11. Electroplating: Test. Dem. 15, 77; J. Chem. Educ. 63 809 (1986); Shak. 4 212-244
12. Current efficiency in electrolysis: J. Chem. Educ. 66 984 (1989)
13. Commercial voltaic cells: J. Chem. Educ. 67 158 (1990)
14. Lead storage battery: Shak. 4 115
15. Corrosion: Test. Dem. 104, 171; J. Chem. Educ. 58 505, 802 (1981), 62 531 (1985), 65 156 (1988)

QUIZZES

Quiz 1

1. How many kilojoules are required to obtain 1.00 g of Al from Al_2O_3, using a voltage of 10.0 V? (1 mol e⁻ = 96480 C)

2. Draw a diagram of a voltaic cell in which the following reaction occurs:

   $Cu(s) + 2Ag^+(aq) \longrightarrow Cu^{2+}(aq) + 2Ag(s)$

   Label anode and cathode and indicate the direction of electron flow.

Quiz 2

1. Using Table 17.1, arrange the reducing agents in the list below in order of increasing strength; do the same with the oxidizing agents.

   NO_3^-, Fe^{2+}, AgBr, SO_4^{2-}, Ag, Zn

2. For the reaction: $Al(s) + 3Ag^+(aq) \longrightarrow Al^{3+}(aq) + 3Ag(s)$, find

   a. E° b. ΔG° c. K

Quiz 3

1. Consider the reaction:

   $Cu(s) + 4H^+(aq) + SO_4^{2-}(aq) \longrightarrow Cu^{2+}(aq) + SO_2(aq) + 2H_2O$

   Referring to Table 17.1, calculate:

   a. E° b. ΔG° c. K

   d. E when P SO_2 = 1 atm, $[Cu^{2+}] = [SO_4^{2-}]$ = 1 M, pH = 3.0

Quiz 4

1. Using Table 17.1, decide whether or not a reaction will occur when a solution of tin(II) bromide is treated with nitric acid. Write a balanced equation for any reaction that occurs.

2. How long will it take to plate 1.00 g of silver from a solution containing the $Ag(CN)_2^-$ ion if a current of 3.25 A is used?

Quiz 5

1. For the reaction: $Co(s) + 2H^+(aq) \longrightarrow Co^{2+}(aq) + H_2(g)$

   calculate, using Table 17.1:

   a. E° b. ΔG°

   c. the pH at which E = 0 when all other species are at standard concentrations.

Answers

Quiz 1 1. 107 kJ

2. Cu anode, Ag cathode; e^- flow from Cu to Ag, cations move to Ag, anions to Cu

Quiz 2 1. reducing agents: Ag < Fe^{2+} < Zn

oxidizing agents: Fe^{2+} < AgBr < SO_4^{2-} < NO_3^-

2. a. +2.48 V   b. -718 kJ   c. 10^{126}

Quiz 3 1. a. -0.184 V   b. +35.5 kJ   c. 5.9×10^{-7} d. -0.539 V

Quiz 4 1. $3Sn^{2+}(aq) + 2NO_3^-(aq) + 8H^+(aq) \longrightarrow 3Sn^{4+}(aq) + 2NO(g) + 4H_2O$

$E° = +0.810$ V

2. 275 s

Quiz 5 1. a. +0.282 V   b. -54.4 kJ   c. 4.77

PROBLEMS

1. a. $2Fe^{3+}(aq) + H_2(g) \longrightarrow 2Fe^{2+}(aq) + 2H^+(aq)$

   b. $Cd(s) + Ni^{2+}(aq) \longrightarrow Cd^{2+}(aq) + Ni(s)$

   c. $2MnO_4^-(aq) + 16H^+(aq) + 10Cl^-(aq) \longrightarrow 2Mn^{2+}(aq) + 5Cl_2(g) + 8H_2O$

3. a. Sn anode, Ag cathode; electrons move from Sn to Ag. Cations flow from Sn half-cell to Ag half-cell; anions flow in opposite direction.

   b. Pt anode surrounded by $H_2(g)$, H^+ ions; Pt cathode in pool of Hg, surrounded by solution saturated with Hg_2Cl_2. Electrons move from anode to cathode. Cations flow through salt bridge from anode to cathode; anions flow in opposite direction.

   c. See Figure 17.14. Pb anode, PbO_2 cathode. Electrons flow from Pb to PbO_2. Cations move to cathode, anions to anode.

5. anode: $Mn(s) \longrightarrow Mn^{2+}(aq) + 2e^-$

   cathode: $Cr^{3+}(aq) + 3e^- \longrightarrow Cr(s)$

   $3Mn(s) + 2Cr^{3+}(aq) \longrightarrow 3Mn^{2+}(aq) + 2Cr(s)$

   $Mn \mid Mn^{2+} \parallel Cr^{3+} \mid Cr$

7. a. H_2O_2 b. MnO_4^- c. Ag^+ d. acidic

9. $PbSO_4 < Br^- < H_2S < Co < Zn$

11. oxid. agents: $Fe^{2+} < Ni^{2+} < H^+ < Cu^+$

    red. agents: $Fe^{2+} < Cu^+ < Zn$

13. a. Sn, Ag, Ni, Co

    b. Cr^{3+}, Cd^{2+}, $PbSO_4$, Tl^+

    c. Cl_2, Cl^-, Cr^{3+}, Mn^{2+}, H_2O, Br^-

15. a. E° = +0.127 V + 0.799 V = +0.926 V
    b. E° = +1.229 V - 0.769 V = +0.460 V
    c. E° = +0.762 V - 0.402 V = +0.360 V

17. a. E° = +0.408 V + 0.154 V = +0.562 V
    b. E° = -1.229 V + 1.763 V = +0.534 V
    c. E° = +0.891 V + 0.401 V = +1.292 V

19. a. E° = +1.68 V + 0.964 V = +2.64 V
    b. E° = +0.408 V + 1.229 V = +1.637 V
    c. E° = -0.339 V + 0.534 V = +0.195 V

21. a. -0.500 V   b. +0.839 V   c. +1.101 V

23. a. E° = +1.001 V - 0.769 V = +0.232 V; spontaneous
    b. E° = +0.762 V + 0.769 V = +1.531 V; spontaneous
    c. E° = -0.339 V + 0.000 V = -0.339 V; nonspontaneous

25. a. E° = +1.512 V - 1.360 V = +0.152 V; yes
    b. E° = +1.33 V - 1.360 V = -0.03 V; no
    c. E° = +0.964 V - 1.360 V = -0.396 V; no

27. a. no reaction: E° = -0.543 V
    b. no reaction; E° = -0.283 V
    c. $2Cr(s) + 3Ni^{2+}(aq) \rightarrow 2Cr^{3+}(aq) + 3Ni(s)$; E° = +0.508 V

29. a. no reaction; E° = -1.498 V
    b. reacts: E° = +0.409 V
    c. no reaction: E° = -0.339 V
    d. reacts; E° = +0.127 V

31. a. possible reductions: $Co^{2+} \rightarrow Co$; E°_{red} = -0.282 V
    $Fe^{2+} \rightarrow Fe$; E°_{red} = -0.409 V

possible oxidations: $Co^{2+} \rightarrow Co^{3+}$; $E^\circ_{ox} = -1.953$ V

$Co \rightarrow Co^{2+}$; $E^\circ_{ox} = +0.282$ V

$Fe^{2+} \rightarrow Fe^{3+}$; $E^\circ_{ox} = -0.769$ V

no reaction

b. possible reductions: $Fe^{2+} \rightarrow Fe$; $E^\circ_{red} = -0.409$ V

$Fe^{3+} \rightarrow Fe^{2+}$ $E^\circ_{red} = +0.769$ V

$Ag^{+} \rightarrow Ag$: $E^\circ_{red} = +0.799$ V

possible oxidation: $Fe^{2+} \rightarrow Fe^{3+}$; $E^\circ_{ox} = -0.769$ V

$Ag^{+}(aq) + Fe^{2+}(aq) \longrightarrow Ag(s) + Fe^{3+}(aq)$

c. possible reductions: $ClO_3^{-} \rightarrow Cl_2$; $E^\circ_{red} = +1.458$ V

$H^{+} \rightarrow H_2$; $E^\circ_{red} = 0.000$ V

possible oxidation: $Hg \rightarrow Hg_2^{2+}$; $E^\circ_{ox} = -0.796$ V

$2ClO_3^{-}(aq) + 12H^{+}(aq) + 10Hg(l) \longrightarrow Cl_2(g) + 5Hg_2^{2+}(aq) + 6H_2O$

33. a. $\Delta G^\circ = 0$; $K = 1$

b. $\Delta G^\circ = -3(96480)(0.500)\text{ J} = -1.45 \times 10^5\text{ J} = -145\text{ kJ}$

$$\ln K = \frac{+1.45 \times 10^5\text{ J}}{(8.31\text{ J/K})(298\text{ K})} = +58.4;\ K = 2 \times 10^{25}$$

c. $\Delta G^\circ = +145\text{ kJ}$; $K = 4 \times 10^{-26}$

35. a. $E^\circ = \frac{+38,700}{96,480} = +0.401$ V

b. +0.201 V c. +0.100 V

37. $E^\circ = +0.004\text{ V} + 0.547\text{ V} = +0.551\text{ V}$

$\Delta G^\circ = -2(9.648 \times 10^4)(+0.551)\text{ J} = -1.06 \times 10^5\text{ J} = -106\text{ kJ}$

$$\ln K = \frac{1.06 \times 10^5}{8.31 \times 298} = +42.9;\ K = 4 \times 10^{18}$$

39. a. $\Delta G^\circ = -2(9.648 \times 10^4)(0.926)\text{ J} = -1.79 \times 10^5\text{ J} = -179\text{ kJ}$

b. $\Delta G^\circ = -4(9.648 \times 10^4)(0.460)\text{ J} = -1.78 \times 10^5\text{ J} = -178\text{ kJ}$

c. $\Delta G^\circ = -2(9.648 \times 10^4)(0.360)\text{ J} = -6.95 \times 10^4\text{ J} = -69.5\text{ kJ}$

41. a. $\ln K = 2(+0.562)/0.0257$ $K = 1 \times 10^{19}$

b. $\ln K = 2(+0.534)/0.0257$ $K = 1 \times 10^{18}$

c. $\ln K = 4(+1.292)/0.0257$ $K = 2 \times 10^{87}$

43. a. $E = E° - \frac{0.0257}{6} \ln \frac{(P\ H_2)^3 \times [Cr^{3+}]^2}{[H^+]^6}$

b. $E° = E°_{ox}\ Cr = +0.744\ V$

c. $E = +0.744\ V - \frac{0.0257}{6} \ln \frac{1.00}{(1.00 \times 10^{-3})^6} = +0.566\ V$

45. $Cr_2O_7^{2-}(aq) + 14H^+(aq) + 6e^- \longrightarrow 2Cr^{3+}(aq) + 7H_2O$

$E_{red} = 1.33\ V - \frac{0.0257}{6} \ln \frac{1(0.100)^2}{(1.00 \times 10^{-4})^{14} x0.100} = +0.79\ V$

47. a. $E = +0.360\ V - \frac{0.0257}{2} \ln \frac{0.50}{0.020} = +0.319\ V$; spontaneous

b. $E = -0.339\ V - \frac{0.0257}{2} \ln \frac{0.0010}{(0.010)^2} = -0.369\ V$; nonspontan.

49. a. $E° = +1.229\ V - 1.360\ V = -0.131\ V$

b. $0 = -0.131 - \frac{0.0257}{4} \ln (P\ Cl_2)^2$; $P\ Cl_2 = 4 \times 10^{-5}$ atm

51. $0.500 = 0.762 - \frac{0.0257}{2} \ln 1/[H^+]^2$

$[H^+] = 3.7 \times 10^{-5}$ M; pH = 4.43

53. a. $E° = +0.127\ V$

b. $0.210 = 0.127 - \frac{0.0257}{2} \ln[Pb^{2+}]$

$[Pb^{2+}] = 1.6 \times 10^{-3}$ M

c. $K_{sp} = [Pb^{2+}] \times [Cl^-]^2 = 1.6 \times 10^{-5}$

55. a. $2.6 \times 10^{22}\ e^- \times \frac{9.648 \times 10^4\ C}{6.022 \times 10^{23}\ e^-} = 4.2 \times 10^3\ C$

b. $m\ H_2 = 0.0432\ mol\ e^- \times \frac{2.016\ g\ H_2}{2\ mol\ e^-} = 0.044\ g\ H_2$

$m\ Cl_2 = 0.0432\ mol\ e^- \times \frac{70.90\ g}{2\ mol\ e^-} = 1.5\ g\ Cl_2$

$m\ OH^- = 0.0432\ mol\ e^- \times \frac{17.01\ g}{1\ mol\ e^-} = 0.73\ g\ OH^-$

57. a. $n\ e^- = (0.400)(7200)C \times \dfrac{1 \text{ mol } e^-}{9.648 \times 10^4 \text{ C}} = 0.0299 \text{ mol } e^-$

$m\ Ag = 0.0299 \text{ mol } e^- \times \dfrac{1 \text{ mol Ag}}{1 \text{ mol } e^-} \times \dfrac{107.9 \text{ g Ag}}{1 \text{ mol Ag}} = 3.22 \text{ g Ag}$

b. $t = 3.22 \text{ g} \times \dfrac{1 \text{ cm}^3}{10.5 \text{ g}} \times \dfrac{1}{4.00 \text{ cm}^2} = 0.0767 \text{ cm}$

59. a. $2.00 \dfrac{C}{s} \times 3600 \text{ s} \times \dfrac{1 \text{ mol } e^-}{9.648 \times 10^4 \text{ C}} \times \dfrac{1 \text{ mol Pb}}{2 \text{ mol } e^-} \times \dfrac{207.2 \text{ g Pb}}{1 \text{ mol Pb}}$

$= 7.73 \text{ g Pb}$

b. $\text{energy} = 7200 \text{ C} \times 12.0 \text{ V} = 8.64 \times 10^4 \text{ J} = 0.0240 \text{ kWh}$

61. Forms tightly adherent coating of Al_2O_3

63. a. requires water to conduct current

b. sets up O_2 concentration cell

65. $Cu^{2+}(aq) + 2Cl^-(aq) \longrightarrow Cu(s) + Cl_2(g)$

$E° = +0.339 \text{ V} - 1.360 \text{ V} = -1.021 \text{ V}$; requires 1.021 V

$Q = 1.00 \text{ g Cu} \times \dfrac{1 \text{ mol Cu}}{63.55 \text{ g Cu}} \times \dfrac{2 \text{ mol } e^-}{1 \text{ mol Cu}} \times \dfrac{9.648 \times 10^4 \text{ C}}{1 \text{ mol } e^-}$

$= 3.04 \times 10^3 \text{ C}$

$\text{energy} = (3.04 \times 10^3)(1.50)\text{J} = 4.56 \times 10^3 \text{ J} = 4.56 \text{ kJ}$

67. a. forms $PbSO_4$ deposit on electrodes

b. conducts current without short-circuiting cell

c. NO_3^- is a powerful oxidizing agent

69. a. $2.00 \text{ A} \times 6.00 \times 10^2 \text{ s} \times \dfrac{1 \text{ mol } e^-}{9.648 \times 10^4 \text{ C}} \times \dfrac{1 \text{ mol Zn}}{2 \text{ mol } e^-} \times \dfrac{65.39 \text{ g Zn}}{1 \text{ mol Zn}}$

$= 0.407 \text{ g}$

b. $n\ NH_3 = \dfrac{2 \times 6.00 \times 10^2 \text{ mol } e^-}{9.648 \times 10^4} \times \dfrac{1 \text{ mol Zn}}{2 \text{ mol } e^-} \times \dfrac{2 \text{ mol } NH_3}{1 \text{ mol Zn}}$

$= 0.0124 \text{ mol}$

$V = \dfrac{(0.0124 \text{ mol})(0.0821 \text{ L·atm/mol·K})(298 \text{ K})}{1.00 \text{ atm}} = 0.304 \text{ L}$

71. $2Ag^+(aq) + Cu(s) \longrightarrow Cu^{2+}(aq) + 2Ag(s)$

$$E = E^\circ - \frac{0.0257}{2} \ln [Cu^{2+}]/[Ag^+]^2$$

a. $\Delta E = \frac{-0.0257}{2} \ln(1.00 \times 10^{-3}) = +0.0888 \text{ V}$

b. $\Delta E = \frac{-0.0257}{2} \ln 1/(1.8 \times 10^{-10})^2 = -0.576 \text{ V}$

c. no effect

73. $AgSCN(s) + e^- \longrightarrow Ag(s) + SCN^-(aq)$ $\quad E^\circ_{red} = +0.0895 \text{ V}$

$Ag(s) \longrightarrow Ag^+(aq) + e^-$ $\quad E^\circ_{ox} = -0.799 \text{ V}$

$AgSCN(s) \longrightarrow Ag^+(aq) + SCN^-(aq)$ $\quad E^\circ = -0.710 \text{ V}$

$\ln K = -0.710/0.0257; \quad K = 1 \times 10^{-12}$

75. Consider one liter of H_2SO_4

$n\ H_2SO_4 = 1286 \text{ g} \times 0.38 \times \frac{1 \text{ mol}}{98.09 \text{ g}} = 5.0 \text{ mol}$

$[H^+] = 5.0 \text{ M}; \quad [SO_4^{2-}] = 0.010 \text{ M}$

$Pb(s) + PbO_2(s) + 4H^+(aq) + 2SO_4^{2-}(aq) \longrightarrow 2PbSO_4(s) + 2H_2O$

$$E = 2.043 \text{ V} - \frac{0.0257}{2} \ln \frac{1}{(5.0)^4(0.010)^2} = 2.007 \text{ V}$$

76. a. $E^\circ = 0.621 \text{ V}$

b. $[Zn^{2+}]$ increases, $[Sn^{2+}]$ decreases

c. $0 = 0.621 - \frac{0.0257}{2} \ln R \qquad R = 1 \times 10^{21}$

d. $[Zn^{2+}] = 2.0 \text{ M}; \ [Sn^{2+}] = 2 \times 10^{-21} \text{ M}$

77. a. $\Delta G^\circ = 2(9.648 \times 10^4)(0.581)\text{J} = 1.12 \times 10^5 \text{ J}$

$\Delta G^\circ = 2(9.648 \times 10^4)(0.197)\text{J} = 0.38 \times 10^5 \text{ J}$

$\Delta G^\circ_{tot} = 1.50 \times 10^5 \text{ J}$

b. $E^{\circ\prime} = -1.50 \times 10^5/(4 \times 9.648 \times 10^4) = -0.389 \text{ V}$

78. anode: $H_2(g) \longrightarrow 2H^+(aq) + 2e^-$

$$E_{ox} = 0.00 \text{ V} - \frac{0.0257}{2} \ln (1 \times 10^{-7})^2 = +0.414 \text{ V}$$

cathode: $2H^+(aq) + 2e^- \longrightarrow H_2(g)$

$E_{red} = 0; \quad E = +0.414 \text{ V}$

CHAPTER 18
Nuclear Reactions

LECTURE NOTES

This material typically is covered near the end of the school year. It serves to hold the interest of restless students. More than any other topic covered in general chemistry, nuclear reactions relate directly to issues that are meaningful to students.

The principles presented in Chapter 18 are readily covered in two lectures. If more time is available, it can profitably be spent in a more detailed discussion of such topics as the effect of radiation on human beings, the pros and cons of nuclear reactors, and the prospects for fusion reactors.

LECTURE 1

I Radioactivity

A. Natural: three different types of radiation

1. Alpha: He-4 nuclei released. Atomic number decreases by 2 units, mass number by 4.

$$^{238}_{92}U \longrightarrow ^{4}_{2}He + ^{234}_{90}Th$$

2. Beta: electron emitted. Atomic number increases by 1 unit, mass number unchanged.

$$^{234}_{90}Th \longrightarrow ^{0}_{-1}e + ^{234}_{91}Pa$$

3. Gamma: high energy radiation. No change in atomic or mass number.

Radioactive series: U-238 decays by a series of steps to form Pb-206:

$$^{238}_{92}U \longrightarrow ^{206}_{82}Pb + ?\ ^{4}_{2}He + ?\ ^{0}_{-1}e$$

Must be 8 alphas and 6 betas to balance the equation

B. Induced radioactivity. Stable nucleus is bombarded by high energy particle (neutron, alpha, etc.)$\rightarrow$ Unstable nucleus

$$^{27}_{13}Al + ^{1}_{0}n \longrightarrow ^{28}_{13}Al \longrightarrow ^{28}_{14}Si + ^{0}_{-1}e$$

If isotope formed has too few neutrons, can get positron decay or K-electron capture.

$$^{11}_{6}C \longrightarrow ^{11}_{5}B + ^{0}_{1}e \quad \text{or} \quad ^{11}_{6}C + ^{0}_{-1}e \longrightarrow ^{11}_{5}B$$

II Rate of Radioactive Decay (First Order)

A. rate = kN

N can be expressed in number of atoms; rate is often expressed in number of atoms decaying per second, or in curies: 1 Ci = 3.700×10^{10} atoms decaying per second

For radium, k = 1.37×10^{-11}/s. Calculate the activity of a one milligram sample of radium.

$N = 1.00 \times 10^{-3} \text{ g} \times \frac{1 \text{ mol}}{226 \text{ g}} \times 6.022 \times 10^{23}$ atoms/mol

$= 2.66 \times 10^{18}$ atoms

$\text{rate} = \frac{1.37 \times 10^{-11}}{\text{s}} \times 2.66 \times 10^{18} \text{ atoms} = 3.64 \times 10^{7}$ atoms/s

$\approx$ 1 millicurie

B. $\ln X_o/X = kt$; $k = 0.693/t_{\frac{1}{2}}$

Half-life of U-238 is 4.5×10^{9} yr. How long does it take for 60% of a U-238 sample to decay?

$k = 0.693/(4.5 \times 10^{9} \text{ yr}) = 1.5 \times 10^{-10}$/yr

$X = 0.40\ X_o$; $X_o/X = 1.00/0.40$

$\ln \frac{1.00}{0.40} = (1.5 \times 10^{-10})t$; $t = 6.1 \times 10^{9}$ yr

C. Age of organic material; measure C-14 content

$$^{14}_{7}N + ^{1}_{0}n \longrightarrow ^{14}_{6}C + ^{1}_{1}H$$

$$^{14}_{6}C \longrightarrow ^{14}_{7}N + ^{0}_{-1}e \quad ; \ t_{\frac{1}{2}} = 5720 \text{ yr}$$

In a living plant or animal, these two processes are in equilibrium, C-14 content is constant. When plant or animal dies, first process stops, C-14 content declines.

Suppose fragment of Shroud of Turin showed C-14 content 0.930 times that of living plant. Age of shroud?

$k = 0.693/5720 \text{ yr} = 1.21 \times 10^{-4}$ yr

$\ln 1/0.930 = (1.21 \times 10^{-4})t$; $t = 600$ yr

LECTURE 2

I Mass-Energy Relations

A. Relation between ΔE and Δm; $\Delta E = c^2 \Delta m$

ΔE (in joules) $= (3.0 \times 10^8)^2 \ \Delta m$ (in kilograms)

ΔE (in kilojoules) $= 9.0 \times 10^{10} \ \Delta m$ (in grams)

B. Calculations

$$^{239}_{94}Pu \longrightarrow \ ^{4}_{2}He + \ ^{235}_{92}U$$

Δm per mole = 234.9934 g + 4.0015 g -239.0006 g = -0.0057 g

ΔE per mole $= -0.0057 \text{ g} \times 9.0 \times 10^{10} \text{ kJ/g} = -5.1 \times 10^8$ kJ

ΔE per gram $= \dfrac{-5.1 \times 10^8 \text{ kJ}}{239 \text{ g}} = -2.1 \times 10^6$ kJ/g

This compares to a maximum of about 50 kJ/g for ordinary chemical reactions.

C. Mass defect = mass(n + p) - mass nucleus

$^{12}_{6}C$: 6(1.00867 g/mol + 1.00728 g/mol) - 11.99671 g/mol
= 0.09899 g/mpl

Binding energy $= 9.00 \times 10^{10}$ kJ/g x 0.09899 g/mol
$= 8.91 \times 10^9$ kJ/mol

Plot of binding energy/(no. of nucleons) vs mass number gives maximum at 50-90. Fission of very heavy nucleus or fusion of very light nucleus evolves energy.

II Fission

$$^{235}_{92}U + \ ^{1}_{0}n \longrightarrow \ ^{90}_{37}Rb + \ ^{144}_{55}Cs + 2\ ^{1}_{0}n$$

Note that:

1. Many different isotopes are formed.
2. More neutrons are produced than consumed, leading to a chain reaction. In nuclear reactor, excess neutrons are absorbed by cadmium rods.
3. Nuclei produced have too many neutrons and hence are intensely radioactive:

$$^{90}_{37}Rb \longrightarrow \ ^{0}_{-1}e + \ ^{90}_{38}Sr$$

This is the principal danger associated with nuclear reactors. Three Mile Island and Chernobyl.

III Fusion

$2\ {}^{2}_{1}H \longrightarrow {}^{4}_{2}He$; $\Delta m = -0.02560$ g; $\Delta E = -2.3 \times 10^{9}$ kJ

per gram: $\Delta E = \dfrac{-2.3 \times 10^{9}\ kJ}{4.0\ g} = -5.7 \times 10^{8}$ kJ/g

ΔE per gram is about 10 times that for fission, 200 times that for radioactivity. Unfortunately, activation energy is very high, since two deuterons repel each other. T required is of the order of 10^{9}°C.

DEMONSTRATIONS

1. Beta rays: Test. Dem. 21
2. Cloud chamber: Test. Dem. 21, 90
3. Simulation of nuclear chain reaction: Shakhashiri Videotapes, Demonstration 49.

QUIZZES

Quiz 1

1. Write a balanced nuclear equation for:
   a. alpha emission by Ra-226.
   b. beta emission by Th-234.
2. The half-life of C-14 is 5720 yr. What fraction of a C-14 sample is left after 2020 yr?

Quiz 2

1. A certain radioactive series starts with Np-237 and ends with Bi-209. Both alpha and beta particles are evolved in the several steps. Write a balanced nuclear equation for the overall reaction.
2. Calculate, using Table 18.2, ΔE per gram of reactant for

   $${}^{19}_{9}F \longrightarrow {}^{18}_{8}O + {}^{1}_{1}H$$

Quiz 3

1. Complete the following nuclear equations:

   a. ${}^{14}_{7}N + \_\_\_\_ \longrightarrow {}^{17}_{8}O + {}^{1}_{1}H$

   b. ${}^{10}_{5}B + {}^{1}_{1}H \longrightarrow \_\_\_\_$

2. The half-life of U-238 is 4.5×10^{9} yr. Calculate
   a. k in 1/s for the decay of U-238
   b. the activity in curies of a sample containing 1.00×10^{23} uranium atoms.

Quiz 4

1. Complete the following statements:
   a. Emission of an electron converts a ______ in the nucleus to a ______.
   b. Capture of a K-electron by a nucleus converts a ______ to a ______.

2. Calculate ΔE per gram of reactant for the following reaction, using Table 18.2.

$$3\ {}^{2}_{1}H \longrightarrow {}^{6}_{3}Li$$

Quiz 5

1. Consider the reaction: ${}^{226}_{88}Ra \longrightarrow {}^{222}_{86}Rn +$ ______
   a. Complete the equation.
   b. Calculate ΔE, using Table 18.2.
   c. If the half-life of Ra-226 is 1590 yr, calculate the rate constant for the reaction.

Answers

Quiz 1 1. a. ${}^{226}_{88}Ra \longrightarrow {}^{222}_{86}Rn + {}^{4}_{2}He$

b. ${}^{234}_{90}Th \longrightarrow {}^{234}_{91}Pa + {}^{0}_{-1}e$

2. 0.783

Quiz 2 1. ${}^{237}_{93}Np \longrightarrow {}^{209}_{83}Bi + 7\ {}^{4}_{2}He + 4\ {}^{0}_{-1}e$

2. 4.07×10^{7} kJ

Quiz 3 1. a. ${}^{4}_{2}He$ b. ${}^{11}_{6}C$

2. a. 4.9×10^{-18}/s b. 1.3×10^{-5} Ci

Quiz 4 1. a. neutron, proton b. proton, neutron

2. -4.05×10^{8} kJ

Quiz 5 1. a. ${}^{4}_{2}He$ b. -4.8×10^{8} kJ c. 4.36×10^{-4}/yr

PROBLEMS

1. ${}^{210}_{82}Pb \longrightarrow {}^{210}_{83}Bi + {}^{0}_{-1}e$; electron

3. a. ${}^{230}_{90}Th \longrightarrow {}^{4}_{2}He + {}^{226}_{88}Ra$

   b. ${}^{210}_{82}Pb \longrightarrow {}^{0}_{-1}e + {}^{210}_{83}Bi$

c. $2\ ^{12}_{6}C \longrightarrow ^{1}_{0}n + ^{23}_{12}Mg$

d. $^{235}_{92}U + ^{1}_{0}n \longrightarrow ^{140}_{56}Ba + 3\ ^{1}_{0}n + ^{93}_{36}Kr$

5. a. $^{87}_{38}Sr$ b. $^{87}_{38}Sr$

7. a. $^{54}_{26}Fe + ^{4}_{2}He \longrightarrow 2\ ^{1}_{1}H + ^{56}_{26}Fe$

b. $^{96}_{42}Mo + ^{2}_{1}H \longrightarrow ^{1}_{0}n + ^{97}_{43}Tc$

c. $^{40}_{18}Ar + ^{4}_{2}He \longrightarrow ^{43}_{19}K + ^{1}_{1}H$

d. $^{31}_{16}S + ^{1}_{0}n \longrightarrow ^{1}_{1}H + ^{31}_{15}P$

9. a. $^{124}_{52}Te$ b. $^{239}_{93}Np$ c. $^{4}_{2}He$ d. $^{24}_{12}Mg$

11. a. $^{12}_{6}C$ b. $^{19}_{9}F$ c. $^{14}_{7}N$

13. a. Fe b. Ge c. Pd

15. 10.0 mg/16 = 0.625 mg

17. a. $k = 0.693/110\ \text{min} = 6.30 \times 10^{-3}/\text{min}$

b. $\ln 1/0.150 = \dfrac{6.30 \times 10^{-3} t}{\text{min}}$ $t = 301\ \text{min}$

19. $\ln \dfrac{0.075}{0.052} = k(8.0\ \text{h});\ k = 0.0461/\text{h};\ t_{\frac{1}{2}} = 15\ \text{h}$

21. rate = kn_t

$\dfrac{(1.3 \times 10^{5})(3.700 \times 10^{10})}{s} = k\left(\dfrac{1.00}{130.9} \times 6.022 \times 10^{23}\right)$

$k = 1.0 \times 10^{-6}/s;\ t_{\frac{1}{2}} = 6.9 \times 10^{5}\ s = 8.0\ d$

23. $^{237}_{93}Np \longrightarrow ^{4}_{2}He + ^{233}_{91}Pa$

$\text{rate} = \dfrac{0.693}{2.20 \times 10^{6}\ \text{yr}} \times \dfrac{1\ \text{yr}}{365.2\ \text{d}} \times \dfrac{1\ \text{d}}{24\ \text{h}} \times \dfrac{1\ \text{h}}{3600\ \text{s}} \times \dfrac{0.500}{237.0}$

$\times 6.022 \times 10^{23} = \dfrac{1.268 \times 10^{7}}{s} \times \dfrac{1\ \text{mCi}}{3.700 \times 10^{7}/s} = 0.343\ \text{mCi}$

25. $\ln \frac{15.3}{5.0} = \frac{0.693 \times t}{5720 \text{ yr}}$ $t = 9.2 \times 10^3$ yr

27. $\ln \frac{1}{0.59} = \frac{0.693 \times t}{12.3 \text{ yr}}$ $t = 9.4$ yr

29. $\ln \frac{2.10}{1.10} = \frac{0.693 \times t}{4.5 \times 10^9 \text{ yr}}$ $t = 4.2 \times 10^9$ yr

31. Ar-K: $\ln 5.13 = 0.693t/(1.26 \times 10^9 \text{ yr})$ $t = 2.97 \times 10^9$ yr

Pb-U: $\ln 1.66 = 0.693t/(4.5 \times 10^9 \text{ yr})$ $t = 3.3 \times 10^9$ yr

Sr-Rb $\ln 1.049 = 0.693t/(4.8 \times 10^{10} \text{ yr})$ $t = 3.3 \times 10^9$ yr

loss of Ar(g)

33. a. $\Delta m = 4.00150 \text{ g} + 225.9771 \text{ g} - 229.9837 \text{ g} = -0.0051 \text{ g}$

b. $\Delta E = 9.00 \times 10^{10} \text{ kJ/g} \times -0.0051 \text{ g} = -4.6 \times 10^8 \text{ kJ}$

$= (-4.6 \times 10^8 \text{ kJ})/230 = -2.0 \times 10^6 \text{ kJ}$

35. a. ${}^{14}_{6}C \longrightarrow 6\ {}^{1}_{1}H + 8\ {}^{1}_{0}n$

$\Delta m = 6(1.00728 \text{ g/mol}) + 8(1.00867 \text{ g/mol}) - 13.99995 \text{ g/mol}$
$= 0.11309 \text{ g/mol}$

b. $\Delta E = 9.00 \times 10^{10} \text{ kJ/g} \times 0.11309 \text{ g/mol} = 1.02 \times 10^{10} \text{ kJ/mol}$

37. Δm for equation $= 4.00150 \text{ g} + 2(0.000549 \text{ g}) - 4(1.00728 \text{ g})$

$= -0.02652 \text{ g}$

$\Delta E/\text{g H} = \frac{-0.02652 \text{ g} \times 9.00 \times 10^{10} \text{ kJ/g}}{4.029 \text{ g H}} = -5.92 \times 10^8 \text{ kJ/g H}$

39. $(3.9 \times 10^{26} \text{ J/s})/(9.00 \times 10^{13} \text{ J/g}) = 4.3 \times 10^{12} \text{ g/s}$

41. a. Δm per mole $= 88.8913 \text{ g} + 143.8817 \text{ g} + 3(0.000549 \text{ g})$
$+ 2(1.00867 \text{ g}) - 234.9934 \text{ g} = -0.2014 \text{ g}$

ΔE per gram $= \frac{9.00 \times 10^{10} \times -0.2014}{235} = -7.71 \times 10^7 \text{ kg}$

b. mass TNT $= (7.71 \times 10^7 \text{ kJ})/(2.76 \times 10^3 \text{ kJ/kg}) = 2.79 \times 10^4 \text{kg}$

43. $\Delta m = 18.99346 \text{ amu} - 17.99477 \text{ amu} - 1.00728 \text{ amu} = -0.00859 \text{ amu}$

spontaneous

45. a. ${}^{235}_{92}U + {}^{1}_{0}n \longrightarrow {}^{144}_{58}Ce + {}^{87}_{35}Br + {}^{0}_{-1}e + 5\ {}^{1}_{0}n$

b. Δm per mole = 143.8817 g + 86.9028 g + 0.000549 g + 4(1.00867 g) - 234.9934 g = -0.1737 g

$$\Delta E \text{ per gram} = \frac{9.00 \times 10^{10}}{235} \times -0.1737 \text{ kJ} = -6.65 \times 10^{7} \text{ kJ}$$

c. $6.65 \times 10^{4} \text{ kJ} \times \frac{1 \text{ mol}}{37.0 \text{ kJ}} \times \frac{80.05 \text{ g}}{1 \text{ mol}} \times \frac{1 \text{ kg}}{10^{3} \text{ g}} = 144 \text{ kg}$

47. a. 45 rems + 10 rems = 55 rems

b. small decrease in white blood cell count

49. amount of radiation absorbed, type of radiation

51. a. F; atomic number b. F; very rapidly c. F; more energy

53. X_o would be higher; t > 1000 yr

55. $[I^-] = [Ag^+] = \frac{2.50 \times 10^{3}}{1.25 \times 10^{10}} \times 0.050 \text{ M} = 1.0 \times 10^{-8} \text{ M}$

$K_{sp} = (1.0 \times 10^{-8})^2 = 1.0 \times 10^{-16}$

57. $n_t = \frac{(1.00 \text{ atm})(1.00 \times 10^{-3} \text{ L})}{(0.0821 \text{ L}\cdot\text{atm/mol}\cdot\text{K})(273 \text{ K})} \times \frac{12.04 \times 10^{23} \text{ atoms}}{1 \text{ mol}}$

$= 5.37 \times 10^{19}$ tritium atoms

$\text{rate} = \frac{0.693}{3.88 \times 10^{8} \text{s}} \times 5.37 \times 10^{19} \times \frac{1 \text{ Ci}}{3.700 \times 10^{10}/\text{s}} = 2.59 \text{ Ci}$

59. $5.0 \times 10^{5} \text{ g Zr} \times \frac{1 \text{ mol Zr}}{91.22 \text{ g Zr}} \times \frac{2 \text{ mol } H_2}{1 \text{ mol Zr}} = 1.10 \times 10^{4} \text{ mol } H_2$

$P = \frac{(1.10 \times 10^{4} \text{ mol})(0.0821 \text{ L}\cdot\text{atm/mol}\cdot\text{K})(328 \text{ K})}{2.0 \times 10^{4} \text{ L}} = 15 \text{ atm}$

61. Cr^{3+}: $765 \frac{\text{cpm}}{\text{g}} \times \frac{161.98 \text{ g}}{1 \text{ mol}} = 1.24 \times 10^{5}$ cpm/mol

$C_2O_4^{2-}$: $512 \frac{\text{cpm}}{\text{g}} \times \frac{90.04 \text{ g}}{1 \text{ mol}} = 4.61 \times 10^{4}$ cpm/mol

$n\, C_2O_4^{2-}/n\, Cr^{3+} = \frac{235}{314} \times \frac{1.24 \times 10^{5}}{4.61 \times 10^{4}} = 2.01$; $Cr(C_2O_4)_2^-$

63. $^{226}_{88}Ra \longrightarrow {}^{4}_{2}He + {}^{222}_{86}Rn$

Δm per mole = 4.0015 g + 221.9703 g - 225.9771 g = -0.0053 g

ΔE per gram = $(9.00 \times 10^{10})(-0.0053)/(226.0) = -2.11 \times 10^{6}$ kg

$$2Cr(s) + 3Cu^{2+}(aq) \longrightarrow 2Cr^{3+}(aq) + 3Cu(s)$$

$$\Delta G^\circ = -6(96.48 \text{ kJ})(1.083) = -626.9 \text{ kJ}$$

64. $$\frac{20 \times 10^{-12} \text{ Ci}}{1 \text{ L}} \times \frac{3.70 \times 10^{10} \text{ atoms/s}}{1 \text{ Ci}} = 0.74 \text{ atom/s}$$

$$t_{\frac{1}{2}} = 3.82 \text{ d} \times \frac{24 \text{ h}}{1 \text{ d}} \times \frac{3600 \text{ s}}{1 \text{ h}} = 3.30 \times 10^5 \text{ s}$$

$$0.74 \frac{\text{atom}}{\text{s}} = \frac{0.693}{3.30 \times 10^5 \text{s}} \times n_t \text{ ; } n_t = 3.52 \times 10^5 \text{ atoms}$$

$$[Rn] = 3.52 \times 10^5 \frac{\text{atom}}{\text{L}} \times \frac{1 \text{ mol}}{6.022 \times 10^{23}} = 5.8 \times 10^{-19} \text{ mol/L}$$

65. a. $$\text{rate} = \frac{5.5 \times 10^{-11}}{\text{min}} \times 1.000 \text{ g} = 5.5 \times 10^{-11} \text{ g/min}$$

5.5×10^{-10} g

b. Δm per mole = -0.0057 g

$$\Delta E = 9.00 \times 10^{10} \times \frac{0.0057}{239} \times 5.5 \times 10^{-10} = 1.2 \times 10^{-3} \text{ kJ}$$

c. $$\frac{1.2 \text{ J/70 kg}}{0.01 \text{ J/kg}} = 1.7 \text{ rads} = 17 \text{ rems}$$

66. a. $$E = \frac{(8.99 \times 10^9)(1.60 \times 10^{-19})^2}{2 \times 10^{-15}} = 1 \times 10^{-13} \text{ J}$$

b. $$\frac{1.2 \times 10^{-13}}{2} = \frac{2.014 \times 10^{-3}}{6.022 \times 10^{23}} \times \frac{1}{2} \times v$$

$$v = [(1.2 \times 10^{-13})(6.022 \times 10^{23})/(2.014 \times 10^{-3})]^{\frac{1}{2}}$$

$$= 6 \times 10^6 \text{ m/s}$$

67. a. Δm = 4.0015 g - 2(2.01355 g) = -0.02560 g

$$\Delta E = \frac{9.00 \times 10^{10} \times -0.02560}{4.027} = -5.72 \times 10^8 \text{ kJ/g}$$

b. $$\Delta E = 5.72 \times 10^8 \frac{\text{kJ}}{\text{g}} \times 1.3 \times 10^{24} \text{ g} \times 1.7 \times 10^{-5}$$

$$= 1.3 \times 10^{28} \text{ kJ}$$

c. $(2.3 \times 10^{17})/(1.3 \times 10^{28}) = 1.8 \times 10^{-11}$

CHAPTER 19
Chemistry of the Metals

LECTURE NOTES

This is a descriptive chapter which concentrates upon the chemistry of the Group 1 and Group 2 metals on the one hand and the transition metals on the other. (Recall that complex ions of the transition metals were covered in Chapter 15). The chapter concludes with two optional topics. One of these is qualitative analysis (Section 19.4) which you may wish to assign as reading material when your students carry out cation analysis in the laboratory. The other deals with alloys, introduced in an effort to pique student interest.

Unless you want to spend a lot of time on Sections 19.4 and 19.5, this chapter is readily covered in two lectures.

LECTURE 1

I Metallurgy

A. Chloride ores; convert to metal by electrolysis

$$NaCl(l) \longrightarrow Na(l) + \frac{1}{2}Cl_2(g)$$

Volume of chlorine at 25°C, 1.00 atm produced by current of 25.0 A in one hour?

$$n\ Cl_2 = (25.0)(3600)C \times \frac{1\ mol\ e^-}{9.648 \times 10^4\ C} \times \frac{\frac{1}{2}\ mol\ Cl_2}{1\ mol\ e^-} = 0.466\ mol$$

V (from ideal gas law) = 11.4 L

B. Oxide ores; most often reduce with CO

Metallurgy of iron:

$$Fe_2O_3(s) + 3CO(g) \longrightarrow 2Fe(s) + 3CO_2(g)$$

$\Delta H° = -26.8$ kJ $\Delta S° = +11.5$ J/K

Spontaneous at all temperatures; carried out at high T to make reaction go quickly. Limestone added to make slag of $CaSiO_3$ with SiO_2 impurity.

C. Sulfide ores; roast in air to form metal or oxide

$HgS(s) + O_2(g) \longrightarrow Hg(g) + SO_2(g)$

$2ZnS(s) + 3\ O_2(g) \longrightarrow 2ZnO(s) + 2SO_2(g)$

Metallurgy of copper; Cu_2S concentrated by flotation, roasted to copper, purified by electrolysis.

II Reactions of Group 1, Group 2 Metals

A. Reaction with hydrogen; form metal hydrides (H^- ion)

$2Na(s) + H_2(g) \longrightarrow 2NaH(s)$

$Ca(s) + H_2(g) \longrightarrow CaH_2(s)$

B. Reaction with water; evolve H_2, form OH^- ions

$2Na(s) + 2H_2O(l) \longrightarrow H_2(g) + 2Na^+(aq) + 2\ OH^-(aq)$

$Ca(s) + 2H_2O(l) \longrightarrow H_2(g) + Ca^{2+}(aq) + 2\ OH^-(aq)$

$\Delta H° = -542.8$ kJ $- 460.0$ kJ $+ 571.6$ kJ $= -431.2$ kJ; exothermic

$\Delta G° = -393.6$ kJ; spontaneous at 25°C

C. Reaction with oxygen

1. Lithium and Group 2 metals yield normal oxides (O^{2-})
2. Na, Ba yield peroxides (O_2^{2-} ion)
3. K, Rb, Cs yield superoxides (O_2^- ion)

Write balanced equations for reaction of O_2 with Mg, Na, K

$2Mg(s) + O_2(g) \longrightarrow 2MgO(s)$

$2Na(s) + O_2(g) \longrightarrow Na_2O_2(s)$

$K(s) + O_2(g) \longrightarrow KO_2(s)$

LECTURE 2

I Redox Chemistry of the Transition Metals

A. Reaction with acid

1. $Zn(s) + 2H^+(aq) \longrightarrow Zn^{2+}(aq) + H_2(g)$
   occurs if $E°_{ox} > 0$
2. $3Cu(s) + 2NO_3^-(aq) + 8H^+(aq) \longrightarrow 3Cu^{2+}(aq) + 2NO(g) + 4H_2O$
   occurs if $E°_{ox} > -0.964$ V

B. Cations may be unstable in water because of

1. Reaction with water

$$Co^{3+}(aq) + e^- \longrightarrow Co^{2+}(aq) \qquad E^\circ_{red} = +1.953\ V$$

$$H_2O \longrightarrow \tfrac{1}{2}\, O_2(g) + 2H^+(aq) + 2e^- \qquad E^\circ_{ox} = -1.229\ V$$

$$2Co^{3+}(aq) + H_2O \longrightarrow 2Co^{2+}(aq) + \tfrac{1}{2}O_2(g) + 2H^+(aq) \qquad E^\circ = +0.724\ V$$

2. Disproportionation

$$Cu^+(aq) + e^- \longrightarrow Cu(s) \qquad E^\circ_{red} = +0.518\ V$$

$$Cu^+(aq) \longrightarrow Cu^{2+}(aq) + e^- \qquad E^\circ_{ox} = -0.161\ V$$

$$2Cu^+(aq) \longrightarrow Cu^{2+}(aq) + Cu(s) \qquad +0.357\ V$$

3. Oxidation by dissolved oxygen

$$Fe^{2+}(aq) \longrightarrow Fe^{3+}(aq) + e^- \qquad E^\circ_{ox} = -0.769\ V$$

$$\tfrac{1}{2}O_2(g) + 2H^+(aq) + 2e^- \longrightarrow H_2O \qquad E^\circ_{red} = +1.229\ V$$

$$2Fe^{2+}(aq) + \tfrac{1}{2}O_2(g) + 2H^+(aq) \longrightarrow 2Fe^{3+}(aq) + H_2O \qquad E^\circ = +0.460\ V$$

C. Oxyanions (CrO_4^{2-}, $Cr_2O_7^{2-}$, MnO_4^-)

1. $2CrO_4^{2-}(aq) + 2H^+(aq) \rightleftharpoons Cr_2O_7^{2-}(aq) + H_2O$

   yellow (chromate); red (dichromate)

   Dichromate ion is stable in acid, chromate stable in basic or neutral solution.

2. Both $Cr_2O_7^{2-}$ and MnO_4^- are powerful oxidizing agents in acidic solution.

$$Cr_2O_7^{2-}(aq) + 14H^+(aq) + 6e^- \longrightarrow 2Cr^{3+}(aq) + 7H_2O$$

$$E_{red} = 1.33\ V - 0.0257\ \ln \frac{[Cr^{3+}]^2}{[Cr_2O_7^{2-}] \times [H^+]^{14}}$$

If $[Cr^{3+}] = [Cr_2O_7^{2-}] = 1.0\ M$, then:

$$E_{red} = 1.33\ V - 0.14\ pH$$

E_{red} decreases from 1.33 V in 1 M acid to +0.35 V in neutral solution.

II Alloys

Phase diagrams; solutions (Fig. 19.16a) and crude mixtures (Fig. 19.16b)

DEMONSTRATIONS

1. Roasting sulfide ores: Test. Dem. 25
2. Preparation of alkali metals: Test. Dem. 193; J. Chem. Educ. 66 438 (1989)
*3. Copper, silver, and gold: Shak. 4 263
4. Reaction of sodium with chlorine: Test. Dem. 146, 200; Shak. 1 61
5. Reaction of sodium with water: Test. Dem. 5; J. Chem. Educ. 58 506 (1981), 61 635 (1984), 69 418 (1992)
6. Reaction of magnesium with steam: Test. Dem. 9, 58, 127, 142
7. Reaction of magnesium with carbon dioxide: Test. Dem. 29; J. Chem. Educ. 55 450 (1978); Shak. 1 90
8. Reaction of metals with hydrochloric acid: Shak. 1 25
9. Equilibrium between Cu(1) and Cu(II): J. Chem. Educ. 50 A59 (1973)
10. Equilibrium between chromate and dichromate: Test. Dem. 182
11. Decomposition of ammonium dichromate: Test. Dem. 5, 53; J. Chem. Educ. 61 908 (1984); Shak. 1 81
12. Oxidation states of manganese: Test. Dem. 175; J. Chem. Educ. 54 302 (1977), 64 624 (1987), 65 451 (1988)

* See also Shakhashiri Videotapes, Demonstration 11

QUIZZES

Quiz 1

1. Write a balanced redox equation for the reaction of silver with nitric acid to form NO(g).
2. Write a balanced equation for the
   a. reaction of sodium with oxygen
   b. reaction of calcium with water
   c. disproportionation of Cu^{+} in water solution

Quiz 2

1. Given that E°_{red} MnO_4^{-} is +1.512 V for the half reaction:

$MnO_4^-(aq) + 8H^+(aq) + 5e^- \longrightarrow Mn^{2+}(aq) + 4H_2O$

calculate E_{red} when $[MnO_4^-] = [Mn^{2+}] = 1.0$ M, pH = 2.0

2. Write balanced equations for
   a. the reaction of CrO_4^{2-} with H^+ ions
   b. the reaction of sodium with chlorine
   c. the reaction of sodium with water

Quiz 3

1. Taking E°_{red} $Au^{3+} \rightarrow Au^+ = +1.400$ V, E°_{red} $Au^+ \rightarrow Au = +1.695$ V
   calculate K for: $3Au^+(aq) \longrightarrow Au^{3+}(aq) + 2Au(s)$

2. Write a balanced equation to explain why
   a. an ionic solid is formed when sodium reacts with hydrogen
   b. a solution of $K_2Cr_2O_7$ turns yellow when treated with base

Quiz 4

1. Given the information

	ΔH°_f	ΔS°
$Fe_2O_3(s)$	-822.2 kJ/mol	90.0 J/K
$H_2(g)$		130.6
$Fe(s)$		27.2
$H_2O(g)$	-241.8	188.7

   estimate the temperature at which the following reaction becomes spontaneous at 1 atm:

   $Fe_2O_3(s) + 3H_2(g) \longrightarrow 2Fe(s) + 3H_2O(g)$

2. Write a balanced equation for the reaction of
   a. cobalt with hydrochloric acid
   b. lithium with hydrogen
   c. potassium with water

Quiz 5

1. Given that E°_{red} for the following half-reaction is +1.33 V:

   $Cr_2O_7^{2-}(aq) + 14H^+(aq) + 6e^- \longrightarrow 2Cr^{3+}(aq) + 7H_2O$

   find the pH at which E_{red} becomes 1.00 V when all species other than H^+ are at unit concentration.

2. What volume of hydrogen gas at 25°C and 1 atm is required to react with 1.00 g of sodium?

Answers

Quiz 1 1. $3Ag(s) + NO_3^-(aq) + 4H^+(aq) \rightarrow 3Ag^+(aq) + NO(g) + 2H_2O$

2. a. $2Na(s) + O_2(g) \rightarrow Na_2O_2$

b. $Ca(s) + 2H_2O \rightarrow Ca^{2+}(aq) + 2\ OH^-(aq) + H_2(g)$

c. $2Cu^+(aq) \rightarrow Cu^{2+}(aq) + Cu(s)$

Quiz 2 1. 1.323 V

2. a. $2CrO_4^{2-}(aq) + 2H^+(aq) \rightarrow Cr_2O_7^{2-}(aq) + H_2O$

b. $2Na(s) + Cl_2(g) \rightarrow 2NaCl(s)$

c. $2Na(s) + 2H_2O \rightarrow 2Na^+(aq) + 2\ OH^-(aq) + H_2(g)$

Quiz 3 1. 9.6×10^9

2. a. $2Na(s) + H_2(g) \rightarrow 2NaH(s)$

b. $Cr_2O_7^{2-}(aq) + 2\ OH^-(aq) \rightarrow 2CrO_4^{2-}(aq) + H_2O$

Quiz 4 1. 698 K

2. a. $Co(s) + 2H^+(aq) \rightarrow Co^{2+}(aq) + H_2(g)$

b. $2Li(s) + H_2(g) \rightarrow 2LiH(s)$

c. $2K(s) + 2H_2O \rightarrow 2K^+(aq) + 2\ OH^-(aq) + H_2(g)$

Quiz 5 1. 2.4

2. 0.532 L

PROBLEMS

1. $2NaCl(l) \rightarrow 2Na(l) + Cl_2(g)$

$$n\ Cl_2 = 1.00\ g\ Na \times \frac{1\ mol\ Na}{22.99\ g\ Na} \times \frac{1\ mol\ Cl_2}{2\ mol\ Na} = 0.0217\ mol\ Cl_2$$

V (from ideal gas law) = 0.486 L

3. a. $2NiS(s) + 3\ O_2(g) \rightarrow 2NiO(s) + 2SO_2(g)$

b. $NiO(s) + CO(g) \rightarrow Ni(s) + CO_2(g)$

5. $\Delta G° = \Delta G^\circ_f\ CO_2 - \Delta G^\circ_f\ CO - \Delta G^\circ_f\ NiO = -45.5$ kJ; yes

7. a. $4Au(s) + 8CN^-(aq) + O_2(g) + 2H_2O \longrightarrow 4Au(CN)_2^-(aq) + 4\ OH^-(aq)$

b. $Zn(s) + 2Au(CN)_2^-(aq) \longrightarrow Zn(CN)_4^{2-}(aq) + 2Au(s)$

9. $n\ O_2 = 10^6\ g\ Fe_2O_3 \times 0.92 \times \dfrac{1\ mol\ Fe_2O_3}{159.70\ g\ Fe_2O_3} \times \dfrac{3\ mol\ CO}{1\ mol\ Fe_2O_3}$

$\times \dfrac{\frac{1}{2}\ mol\ O_2}{1\ mol\ CO} = 8.64 \times 10^3\ mol\ O_2$

$V\ O_2(g)$ from ideal gas law $= 2.11 \times 10^5$ L

$V\ air = 2.11 \times 10^5\ L\ O_2 \times \dfrac{1\ L\ air}{0.21\ L\ O_2} \times \dfrac{1\ ft^3}{28.32\ L} = 3.6 \times 10^4\ ft^3$

11. $MS(s) + 3/2\ O_2(g) \longrightarrow MO(s) + SO_2(g)$

Let $\mathcal{M}$ = molar mass of M

$\dfrac{\mathcal{M} + 16.00}{\mathcal{M} + 32.07} = \dfrac{2.368}{2.876} = 0.8234$; $\mathcal{M} = 58.95$ g/mol

13. a. Sr_3N_2; strontium nitride

b. $SrBr_2$; strontium bromide

c. $Sr(OH)_2$; strontium hydroxide

d. SrO; strontium oxide

15. a. $Mg(s) + Cl_2(g) \longrightarrow MgCl_2(s)$; magnesium chloride

b. $BaO_2(s) + 2H_2O \longrightarrow H_2O_2(aq) + Ba^{2+}(aq) + 2\ OH^-(aq)$
hydrogen peroxide, barium hydroxide

c. $2Li(s) + S(s) \longrightarrow Li_2S(s)$; lithium sulfide

d. $2Na(s) + 2H_2O \longrightarrow H_2(g) + 2Na^+(aq) + 2\ OH^-(aq)$
hydrogen, sodium hydroxide

17. $CaH_2(s) + 2H_2O \longrightarrow 2H_2(g) + Ca^{2+}(aq) + 2\ OH^-(aq)$

$n\ H_2 = \dfrac{(1.10\ atm)(25.0\ L)}{(0.0821\ L\cdot atm/mol\cdot K)(298\ K)} = 1.12\ mol$

$m = 1.12\ mol\ H_2 \times \dfrac{1\ mol\ CaH_2}{2\ mol\ H_2} \times \dfrac{42.10\ g\ CaH_2}{1\ mol\ CaH_2} = 23.6\ g\ CaH_2$

19. a. $2CrO_4^{2-}(aq) + 2H^+(aq) \longrightarrow Cr_2O_7^{2-}(aq) + H_2O$

b. $4MnO_4^-(aq) + 2H_2O \longrightarrow 4MnO_2(s) + 4\ OH^-(aq) + 3\ O_2(g)$

21. $Hg(l) + 4Cl^-(aq) + 2NO_3^-(aq) + 4H^+(aq) \rightarrow HgCl_4^{2-}(aq) + 2NO_2(g) + 2H_2O$

23. a. $Cu(s) + 2NO_3^-(aq) + 4H^+(aq) \longrightarrow Cu^{2+}(aq) + 2NO_2(g) + 2H_2O$

b. $2Cr(OH)_3(s) + 3ClO^-(aq) + 4\ OH^-(aq) \longrightarrow 2CrO_4^{2-}(aq) + 3Cl^- + 5H_2O$

25. a. E° = +0.402 V; spontaneous

b. E° = +0.912 V; spontaneous

c. E° = +0.282 V; spontaneous

d. E° = -0.799 V; nonspontaneous

e. E° = -1.498 V; nonspontaneous

27. Au^+: E° = 1.695 V - 1.400 V = +0.295 V

29. a. $3Fe^{2+}(aq) \longrightarrow 2Fe^{3+}(aq) + Fe(s)$; E° = -1.178 V

$\ln K = -2(1.178)/0.0257$; $K = 2 \times 10^{-40}$

b. $\frac{[Fe^{3+}]^2}{[Fe^{2+}]^3} = 2 \times 10^{-40}$; $[Fe^{3+}]^2 = 2 \times 10^{-43}$; $[Fe^{3+}] = 4 \times 10^{-22} M$

31. I Cl^-, AgCl; II H_2S, Bi_2S_3; III H_2S, CoS;

IV CO_3^{2-}, $MgCO_3$

33. a. K^+ b. Sb^{3+} c. Zn^{2+}

35. NiS; $P = 1 \times 10^{-22} < K_{sp}$; no precipitate

CuS: $P = 1 \times 10^{-22} > K_{sp}$; precipitate

37. a. $Ag^+(aq) + Cl^-(aq) \longrightarrow AgCl(s)$

b. $Cd^{2+}(aq) + H_2S(aq) \longrightarrow CdS(s) + 2H^+(aq)$

c. $Ca^{2+}(aq) + CO_3^{2-}(aq) \longrightarrow CaCO_3(s)$

39. liquid; liquid + solid solution; solid solution; about 1250°C

41. a. n Cu = 75.0 g/(63.55 g/mol) = 1.18 mol

n Ni = 25.0 g/(58.69 g/mol) = 0.426 mol

$X\ Cu = \frac{1.18}{1.18 + 0.426} = 0.735$

b. $0.925 \times 1.00\ g\ Ag \times \frac{1\ mol\ Ag}{107.9\ g\ Ag} \times \frac{6.022 \times 10^{23}\ atoms}{1\ mol\ Ag}$

$= 5.16 \times 10^{21}$ atoms

43. See Table 19.A

45. Forms enzyme involved in functioning of oil glands

47. $2Na(s) + 2H_2O \longrightarrow H_2(g) + 2Na^+(aq) + 2\ OH^-(aq)$

$$n\ H_2 = \frac{(2.73\ L)(752/760\ atm)}{(0.0821\ L\cdot atm/mol\cdot K)(295\ K)} = 0.112\ mol$$

$$m\ Na = 0.112\ mol\ H_2 \times \frac{2\ mol\ Na}{1\ mol\ H_2} \times \frac{22.99\ g\ Na}{1\ mol\ Na} = 5.13\ g\ Na$$

49. $[Pb^{2+}] = 1.7 \times 10^{-5}/(2.0 \times 10^{-1})^2 = 4.2 \times 10^{-4}\ M$

51. $Fe_2O_3(s) + 3H_2(g) \longrightarrow 2Fe(s) + 3H_2O(g)$

$\Delta H° = 3(-241.8\ kJ) + 824.2\ kJ = +98.8\ kJ$

$\Delta S° = +0.1415\ kJ/K$

$T = 98.8/0.1415 = 698\ K = 425°C$

53. $$2.2 \times 10^5\ g\ Ni \times \frac{90.76\ g\ NiS}{58.69\ g\ Ni} = 3.4 \times 10^5\ g\ NiS$$

55. a. $Fe^{2+}(aq) \longrightarrow Fe^{3+}(aq) + e^-$

$MnO_4^-(aq) + 5e^- + 8H^+(aq) \longrightarrow Mn^{2+}(aq) + 4H_2O$

$5Fe^{2+}(aq) + MnO_4^-(aq) + 8H^+(aq) \longrightarrow 5Fe^{3+}(aq) + Mn^{2+}(aq) + 4H_2O$

b. $E° = +0.743\ V$

c. $$m\ Fe = (0.05563 \times 0.0200)mol\ MnO_4^- \times \frac{5\ mol\ Fe}{1\ mol\ MnO_4^-} \times \frac{55.85\ g\ Fe}{1\ mol\ Fe}$$

$= 0.311\ g$

$$\%\ Fe = \frac{0.311}{0.3500} \times 100 = 88.8\%$$

57. Let x = mass BaO; 22.38 - x = mass BaO_2

$$20.00\ g = \frac{137.3}{153.3}x + \frac{137.3}{169.3}(22.38\ g - x)$$

x = 21.86 g; 98% BaO, 2% BaO_2

58. a. $Fe(OH)_3(s) + 3H_2C_2O_4(aq) \longrightarrow Fe(C_2O_4)_3^{3-}(aq) + 3H^+(aq) + 3H_2O$

b. $$1.0\ g\ Fe(OH)_3 \times \frac{1\ mol\ Fe(OH)_3}{106.88\ g\ Fe(OH)_3} \times \frac{3\ mol\ H_2C_2O_4}{1\ mol\ Fe(OH)_3}$$

$= 0.0281 \text{ mol } H_2C_2O_4$

$V = 0.0281 \text{ mol}/(0.10 \text{ mol/L}) = 2.8 \times 10^2 \text{ mL}$

59. $6Fe^{2+}(aq) + Cr_2O_7^{2-}(aq) + 14H^+(aq) \longrightarrow 6Fe^{3+}(aq) + 2Cr^{3+}(aq) + 7H_2O$

$$n\ Fe^{2+} = (0.01350 \times 0.100)\text{mol } Cr_2O_7^{2-} \times \frac{6 \text{ mol } Fe^{2+}}{1 \text{ mol } Cr_2O_7^{2-}}$$

$$= 8.10 \times 10^{-3} \text{ mol } Fe^{2+}$$

$n\ Fe^{2+}$ used $= 0.07500 \times 0.125 - 0.00810 = 0.001275$

$5Fe^{2+}(aq) + MnO_4^-(aq) + 8H^+(aq) \longrightarrow 5Fe^{3+}(aq) + Mn^{2+}(aq) + 4H_2O$

$$n\ MnO_4^- = 1.275 \times 10^{-3} \text{ mol } Fe^{2+} \times \frac{1 \text{ mol } MnO_4^-}{5 \text{ mol } Fe^{2+}}$$

$$= 2.55 \times 10^{-4} \text{ mol } MnO_4^-$$

$$\text{mass Mn} = 2.55 \times 10^{-4} \text{ mol} \times \frac{54.94 \text{ g}}{1 \text{ mol}} = 0.0140 \text{ g}$$

$$\% \text{ Mn} = \frac{0.0140}{0.500} \times 100 = 2.80\%$$

60. $\Delta H° = +520.0$ kJ; $\Delta S° = +0.1840$ kJ/K

$T = 2830 \text{ K} \approx 2560°C$

61. $Cr_2O_7^{2-}(aq) + 2\ OH^-(aq) \longrightarrow 2CrO_4^{2-}(aq) + H_2O$

$2Ag^+(aq) + CrO_4^{2-}(aq) \longrightarrow Ag_2CrO_4(s)$

$Ag_2CrO_4(s) + 4NH_3(aq) \longrightarrow 2Ag(NH_3)_2^+(aq) + CrO_4^{2-}(aq)$

$2Ag(NH_3)_2^+(aq) + CrO_4^{2-}(aq) + 4H^+(aq) \longrightarrow Ag_2CrO_4(s) + 4NH_4^+(aq)$

CHAPTER 20
Chemistry of the Nonmetals

LECTURE NOTES

This is a heavily descriptive chapter that puts a great deal of emphasis on writing equations, particularly net ionic equations. Students probably have more trouble with that topic than with any other in general chemistry. It is certainly true that many of the equations in this chapter have appeared before, albeit in a different context. Remember, though, most of the students had trouble with these equations the first time around, so a little review won't hurt.

As in every descriptive chapter, not every topic needs to be covered in lecture. With that in mind, two lectures should be sufficient for this chapter.

<u>LECTURE 1</u>

I <u>Preparation of the Elements</u>

A. Nitrogen and oxygen from liquid air.

B. Sulfur from underground deposits; Frasch process

C. Halogens

F_2 by electrolysis of HF, Cl_2 by electrolysis of aqueous NaCl (Chapter 17). Bromine and iodine prepared using chlorine.

$Cl_2(g) + 2Br^-(aq) \longrightarrow Br_2(l) + 2Cl^-(aq)$; $E° = +0.283$ V

$Cl_2(g) + 2I^-(aq) \longrightarrow I_2(s) + 2Cl^-(aq)$ $E° = +0.826$ V

II <u>Allotropy</u>

A. O_2 vs O_3. Electronic structures

B. White and red phosphorus; structure and properties

C. Sulfur. Rhombic and monoclinic structures. Behavior of liquid sulfur; polymerization between 160-250°C. Chains:

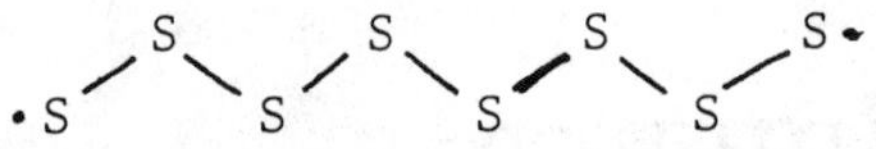

$\cdot S - S - S - S - S - S - S - S \cdot$

III Hydrogen Compounds

A. Ammonia

Bronsted base: $NH_3(aq) + H_2O \rightleftharpoons NH_4^+(aq) + OH^-(aq)$

Lewis base: $2NH_3(aq) + Ag^+(aq) \longrightarrow Ag(NH_3)_2^+(aq)$

Precipitating agent: $Fe^{3+}(aq) + 3NH_3(aq) + 3H_2O \longrightarrow$
$Fe(OH)_3(s) + 3NH_4^+(aq)$

Can also act as reducing agent, but not as oxidizing agent

B. Hydrogen sulfide

Bronsted acid, precipitating agent, reducing agent

C. Hydrogen peroxide

$H_2O_2(aq) + 2H^+(aq) + 2e^- \longrightarrow 2H_2O;\ E^\circ_{red} = +1.763\ V$

$H_2O_2(aq) \longrightarrow O_2(g) + 2H^+(aq) + 2e^-;\ E^\circ_{ox} = -0.695\ V$

strong oxidizing agent, weak reducing agent

D. HF, HCl

Reaction with CO_3^{2-}:

$2H^+(aq) + CO_3^{2-}(aq) \longrightarrow CO_2(g) + H_2O$

$2HF(aq) + CO_3^{2-}(aq) \longrightarrow CO_2(g) + H_2O + 2F^-(aq)$

LECTURE 2

I Oxygen Compounds

A. Molecular structures of oxides of N, P, S

Electronic structure of N_2O_4?

34 valence e^-; skeleton : (O)(O)N - N(O)(O)

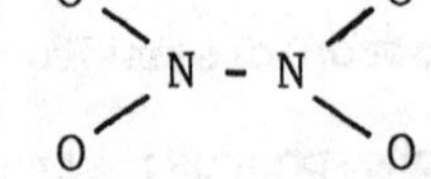

24 e^- left: $:\ddot{O}=N(-\ddot{O}:) - N(-\ddot{O}:)=\ddot{O}:$ many resonance forms

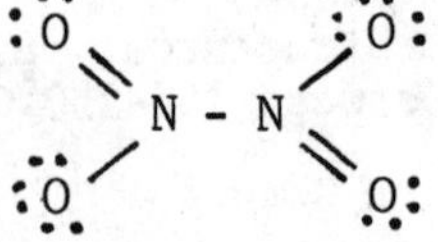

B. Reaction with water

$N_2O_5(g) + H_2O(l) \longrightarrow 2HNO_3(l)$ Note that oxidation state is unchanged

$CO_2(g) + H_2O \longrightarrow H_2CO_3(aq)$

II Oxoacids, Oxoanions

A. Acid strength
Increases with electronegativity and oxidation number of central atom.

$HClO_4$	>	$HClO_3$	>	$HClO_2$	>	$HClO$
very strong		strong		weak		very weak

$HClO > HBrO > HIO$

Rationale: strength depends upon how readily O-H bond is broken to form proton. Increase in electronegativity or oxidation number tends to draw electrons away from O atom bonded to hydrogen, making ionization easier.

B. Oxidizing and reducing strength

1. Species in highest oxidation state (NO_3^-, SO_4^{2-}) can act only as oxidizing agents. In intermediate state (NO_2^-, SO_3^{2-}) can act as either oxidizing or reducing agent.

$NO_2^-(aq) + H_2O \longrightarrow NO_3^-(aq) + 2H^+(aq) + 2e^-$

$NO_2^-(aq) + 2H^+(aq) + e^- \longrightarrow NO(g) + H_2O$

2. Oxidizing strength increases with H^+ ion concentration

$NO_3^-(aq) + 4H^+(aq) + 3e^- \longrightarrow NO(g) + H_2O$

$[H^+]$	1.0 M	1.0×10^{-7} M	1.0×10^{-14} M
E_{red}	+0.964 V	+0.412 V	-0.139 V

C. Nitric acid Strong acid, strong oxidizing agent. Can be reduced to NO_2, NO, or even NH_4^+.

D. Sulfuric acid Strong acid, dehydrating agent (effect on sugar), relatively weak oxidizing agent ($E^\circ_{red} = +0.155$ V).

DEMONSTRATIONS

1. Preparation of chlorine: Test. Dem. 43; Shak. 2 220

2. Reaction of chlorine with bromide, iodide ions: Test. Dem. 5, 213
3. Oxidation states of Br, I: J. Chem. Educ. 64 607 (1987)
4. Allotropes of sulfur: Test. Dem. 41, 101; Shak. 1 243
5. Oxidation of phosphorus: Shak. 1 74, 186
6. Preparation of hydrogen halides: Test. Dem. 44
7. Preparation and properties of HCl: Shak. 2 198
8. Preparation and properties of ammonia: Shak. 2 202
9. Reactions of hydrogen peroxide: Test. Dem. 62, 152, 213
10. Formation of phosphine: Test. Dem. 39, 100
11. Etching glass with HF: Shak. 3 80
12. Oxides of nitrogen: Test. Dem. 99
13. Properties of nitric and sulfuric acids: Shak. 3 70
14. Oxidizing properties of nitric acid: Test. Dem. 99

*15. Reaction of sulfuric acid with sugar: Test. Dem. 35; Shak. 1 77

*16. Decomposition of nitrogen triiodide: Shak. 1 96

* See also Shakhashiri Videotapes, Demonstrations 26, 27

QUIZZES

Quiz 1

1. Write a balanced equation for
   a. the reaction of one mole of phosphoric acid with two moles of OH^- ions.
   b. the reaction of ammonia with Al^{3+} to form a precipitate.
2. Write a balanced redox equation for the disproportionation of $HClO_2$ to ClO_3^- and HClO in acidic solution.

Quiz 2

1. Write balanced equations for
   a. the preparation of chlorine by electrolysis of a water solution of NaCl.
   b. the disproportionation of H_2O_2
2. Write a balanced redox equation for the reaction of hydrogen sulfide with oxygen to form solid sulfur (acid solution).

Quiz 3

1. Give the molecular formulas of
   a. oxygen and ozone b. monoclinic and rhombic sulfur
   c. white phosphorus

2. Write balanced net ionic equations for the precipitation of
   a. $Fe(OH)_3$, using NH_3
   b. $PbCl_2$, using HCl
   c. Bi_2S_3, using H_2S

Quiz 4

1. Write a balanced equation for the reaction of hydrogen sulfide with
   a. Bi^{3+} b. OH^-
   c. dilute nitric acid (products include NO and S)

2. Draw the Lewis structure of N_2O

Quiz 5

1. Taking E°_{red} $HClO \rightarrow Cl_2$ = +1.630 V; E°_{red} Cl_2 = +1.360 V, calculate:
   a. E° for the disproportionation of one mole of Cl_2 in acidic solution.
   b. K for the reaction in (a).

2. Write a balanced equation for the commercial preparation of
   a. $Br_2(l)$ b. $F_2(g)$

Answers

Quiz 1 1.a. $H_3PO_4(aq) + 2\ OH^-(aq) \rightarrow HPO_4^{2-}(aq) + 2H_2O$

b. $Al^{3+}(aq) + 3NH_3(aq) + 3H_2O \rightarrow Al(OH)_3(s) + 3NH_4^+(aq)$

2. $2HClO_2(aq) \rightarrow H^+(aq) + HClO(aq) + ClO_3^-(aq)$

Quiz 2 1.a. $2Cl^-(aq) + 2H_2O \rightarrow Cl_2(g) + H_2(g) + 2\ OH^-(aq)$

b. $2H_2O_2(aq) \rightarrow 2H_2O + O_2(g)$

2. $2H_2S(g) + O_2(g) \rightarrow 2S(s) + 2H_2O$

Quiz 3 1. a. O_2, O_3 b. S_8, S_8 c. P_4

2. a. $Fe^{3+}(aq) + 3NH_3(aq) + 3H_2O \rightarrow Fe(OH)_3(s) + 3NH_4^+(aq)$

b. $Pb^{2+}(aq) + 2Cl^-(aq) \rightarrow PbCl_2(s)$

c. $2Bi^{3+}(aq) + 3H_2S(aq) \rightarrow Bi_2S_3(s) + 6H^+(aq)$

Quiz 4 1. a. $2Bi^{3+}(aq) + 3H_2S(aq) \longrightarrow Bi_2S_3(s) + 6H^+(aq)$

b. $H_2S(aq) + OH^-(aq) \longrightarrow HS^-(aq) + H_2O$

c. $3H_2S(aq) + 2NO_3^-(aq) + 2H^+(aq) \longrightarrow 3S(s) + 2NO(g) + 4H_2O$

2. $:N \equiv N - \ddot{\underset{..}{O}}:$

Quiz 5 1. a. -0.270 V b. 2.7×10^{-5}

2. a. $Cl_2(g) + 2Br^-(aq) \longrightarrow Br_2(l) + 2Cl^-(aq)$

b. $2HF(l) \longrightarrow H_2(g) + F_2(g)$

PROBLEMS

1. a. periodic acid b. bromite ion c. hypoiodous acid
   d. sodium chlorate

3. a. $HClO_3$ b. HIO_4 c. HBrO d. HI

5. a. NH_3 b. H_2S c. NaCl

7. a. N_2O_5 b. N_2O_3 c. SO_3

9. a. NH_3 b. N_2O c. H_2O_2 d. SO_3

11. a. NH_3 b. P_2H_4 c. H_2O_2

13. a. S^{2-} b. HSO_3^-, SO_3^{2-} c. H_2SO_3, H_2SO_4, H_2S

15. a. $2HF(l) \longrightarrow H_2(g) + F_2(g)$

b. $H_2O_2(aq) + 2I^-(aq) + 2H^+(aq) \longrightarrow 2H_2O + I_2(s)$

17. a. $3I_2(s) + 6\ OH^-(aq) \longrightarrow IO_3^-(aq) + 5I^-(aq) + 3H_2O$

b. $4Cl_2(g) + 8\ OH^-(aq) \longrightarrow 7Cl^-(aq) + ClO_4^-(aq) + 4H_2O$

19. a. $Cl_2(g) + 2I^-(aq) \longrightarrow 2Cl^-(aq) + I_2(s)$

b. $F_2(g) + 2Br^-(aq) \longrightarrow 2F^-(aq) + Br_2(l)$

c. N. R.

d. $Br_2(l) + 2I^-(aq) \longrightarrow 2Br^-(aq) + I_2(s)$

21. a. $2HF(l) \longrightarrow H_2(g) + F_2(g)$

b. $Cl_2(g) + 2Br^-(aq) \longrightarrow 2Cl^-(aq) + Br_2(l)$

c. $NH_3(aq) + H^+(aq) \longrightarrow NH_4^+(aq)$

23. a. $Cu^{2+}(aq) + 4NH_3(aq) \longrightarrow Cu(NH_3)_4^{2+}(aq)$

b. $H^+(aq) + NH_3(aq) \longrightarrow NH_4^+(aq)$

c. $Al^{3+}(aq) + 3NH_3(aq) + 3H_2O \longrightarrow Al(OH)_3(s) + 3NH_4^+(aq)$

25. a. $H^+(aq) + OH^-(aq) \longrightarrow H_2O$

b. $Ag(s) + NO_3^-(aq) + 2H^+(aq) \longrightarrow Ag^+(aq) + NO_2(g) + H_2O$

c. $5Cd(s) + 2NO_3^-(aq) + 12H^+(aq) \longrightarrow 5Cd^{2+}(aq) + N_2(g) + 6H_2O$

27. b, c

29. a. O_3 vs O_2 molecules
b. P_4 molecules vs network covalent
c. S_8 molecules vs long chains

31. a. ·N(=O:)(–O:)   b. ·N = O: c. :O – S = O: d. :O – S(=O:) – O:

33. a, b, c

35. a. H – O – N(=O:) – O: b. H – O – S(:O:)(:O:) – O – H   c. H – O – P(:O:)(O – H) – O – H

37. a. H – O – Br(:O:)(:O:) – O: b. H – N(H) – N(H) – H   c. H – O – P(:O:)(O – H) – O – H

39. a. $1.00 \times 10^3 \text{ g Br}^- \times \frac{1 \text{ g seawater}}{65 \times 10^{-6} \text{ g Br}^-} \times \frac{1 \text{ lb}}{453.6 \text{ g}} \times \frac{1 \text{ ft}^3}{64.0 \text{ lb}}$

$= 5.3 \times 10^2 \text{ ft}^3$

b. $n\ Cl_2 = 1.00 \times 10^3 \text{ g Br}^- \times \frac{1 \text{ mol Br}^-}{79.90 \text{ g Br}^-} \times \frac{1 \text{ mol Cl}_2}{2 \text{ mol Br}^-}$

$= 6.26 \text{ mol Cl}_2$

V (ideaL gas law) = 1.50×10^2 L

41. mass I_2 = 0.200 g MnO_2 x $\frac{1 \text{ mol } MnO_2}{86.94 \text{ g } MnO_2}$ x $\frac{1 \text{ mol } I_2}{1 \text{ mol } MnO_2}$ x $\frac{253.8 \text{ g } I_2}{1 \text{ mol } I_2}$

= 0.584 g I_2

43. $NH_4NO_3(s) \longrightarrow N_2(g) + 2H_2O(g) + \frac{1}{2}O_2(g)$

n gas = 1.000×10^3 g NH_4NO_3 x $\frac{1 \text{ mol } NH_4NO_3}{80.05 \text{ g } NH_4NO_3}$ x $\frac{3.5 \text{ mol gas}}{1 \text{ mol } NH_4NO_3}$

= 43.72 mol

P = 2.77×10^3 atm

45. $2NaNO_3(s) \longrightarrow 2NaNO_2(s) + O_2(g)$

n O_2 = $\frac{(0.1250 \text{ L})(731/760 \text{ atm})}{(0.0821 \text{ L·atm/mol·K})(296 \text{ K})}$ = 4.95×10^{-3} mol

m $NaNO_3$ = 4.95×10^{-3} mol O_2 x $\frac{2 \text{ mol } NaNO_3}{1 \text{ mol } O_2}$ x $\frac{85.00 \text{ g } NaNO_3}{1 \text{ mol } NaNO_3}$

= 0.841 g $NaNO_3$

% $NaNO_3$ = $\frac{0.841}{1.500}$ x 100% = 56.1%

47. $[H^+] \times [Br^-] \times [HBrO] = 1.2 \times 10^{-9}$

$[H^+] = (1.2 \times 10^{-9})^{1/3} = 1.1 \times 10^{-3}$ M; pH = 2.97

49. $K = (0.7324)^2/(2.80 \times 10^{-3})^2 = 6.84 \times 10^4$

51. $[Ba^{2+}] \times [F^-]^2 = 1.8 \times 10^{-7}$

$[F^-]^3 = 3.6 \times 10^{-7}$; $[F^-] = 7.1 \times 10^{-3}$ M

53. $\Delta H°$ = -795.4 kJ + 1002.8 kJ = +207.4 kJ

$\Delta S°$ = +0.1161 kJ/K + 0.2862 kJ/K -0.2230 kJ/K - 0.3030 kJ/K

= -0.1237 kJ/K

$\Delta G°$ = 207.4 kJ + 298 K(0.1237 kJ/K) = +244.3 kJ; nonspontaneous

55. $\Delta H°$ = -332.6 kJ + 320.1 kJ = -12.5 kJ

$\Delta G°$ = -8.31×10^{-3} kJ/K x 298 K x $\ln(6.9 \times 10^{-4})$ = +18.0 kJ

$\Delta S°$ = $\frac{-30.5 \text{ kJ}}{298 \text{ K}}$ = $-0.0138 \frac{\text{kJ}}{\text{K}}$ - S° HF(aq)

S° HF(aq) = 0.088 kJ/K

57. a. $\Delta H° = 4(90.2\ kJ) + 6(-241.8\ kJ) - 4(-46.1\ kJ)$

$= -905.6\ kJ$; exothermic

b. positive

$\Delta S° = 6(0.1887\ kJ/K) + 4(+0.2107\ kJ/K) - 5(+0.2050\ kJ/K)$
$-4(0.1923\ kJ/K) = +\ 0.1808\ kJ/K$

c. $\Delta G° = -905.6\ kJ - 298(0.1808\ kJ/K) = -959.5\ kJ$; yes

d. none; $\Delta H°$ and $\Delta S°$ have opposite signs

59. $2I^-(aq) \longrightarrow I_2(s) + 2e^-$

$$Q = 1\ mol\ I_2 \times \frac{2\ mol\ e^-}{1\ mol\ I_2} \times \frac{9.648 \times 10^4\ C}{1\ mol\ e^-} = 1.930 \times 10^5\ C$$

$$E = 1.930 \times 10^5\ C \times 5.00\ V \times \frac{1\ kJ}{10^3\ J} = 965\ kJ$$

61. $Cl^-(aq) + H_2O \longrightarrow ClO^-(aq) + 2H^+(aq) + 2e^-$

$$n\ ClO^- = 1.500 \times 10^6\ g \times 0.0500 \times \frac{1\ mol\ NaClO}{74.44\ g\ NaClO} = 1008\ mol\ ClO^-$$

$n\ e^- = 2016\ mol$

$$t = \frac{2016\ mol\ e^- \times 9.648 \times 10^4\ C/mol\ e^-}{2.00 \times 10^3\ C/s} = 9.72 \times 10^4\ s$$

63. a. $E° = +0.63\ V$; yes b. $E° = -1.104\ V$; no

c. $E° = -0.161\ V$; no d. $E° = +0.382\ V$; yes

65. $2NO_3^-(aq) + 3SO_2(g) + 2H_2O \longrightarrow 2NO(g) + 3SO_4^{2-}(aq) + 4H^+(aq)$

$$E = +0.809\ V - \frac{0.0257}{6} \ln \frac{(0.100)^3(5.0 \times 10^{-5})^4}{(0.100)^2} = +0.988\ V$$

67. See p. 571

69. See p. 572

71. a. $HClO$ b. HIO_4 c. $HBrO_4$

73. $SiO_2(s) + 4HF(aq) \longrightarrow SiF_4(g) + 2H_2O$

$$n\ HF = 1.00\ g\ SiO_2 \times \frac{1\ mol\ SiO_2}{60.09\ g\ SiO_2} \times \frac{4\ mol\ HF}{1\ mol\ SiO_2} = 0.0666\ mol\ HF$$

$V = 0.0666\ mol/(2.0\ mol/L) = 33.3\ mL$

75. $\mathcal{M} = \frac{dRT}{P} = \frac{(0.8012\ g/L)(0.0821\ L\cdot atm/mol\cdot K)(973\ K)}{1.00\ atm}$

$\mathcal{M} = 64.0\ g/mol;\ S_2$

77. contaminated by NO_2

79. depth of deposit, purity of S, density of S, molar masses of S, H_2SO_4

80. Consider one kilogram of quartz

value of gold = $0.010\ g \times \frac{1\ oz}{31.1\ g} \times \frac{\$425}{1\ oz} = \$0.14$

cost of HF = $1.000 \times 10^3\ g\ SiO_2 \times \frac{1\ mol\ SiO_2}{60.09\ g\ SiO_2} \times \frac{4\ mol\ HF}{1\ mol\ SiO_2}$

$\times \frac{20.01\ g\ HF}{1\ mol\ HF} \times \frac{1\ g\ acid}{0.50\ g\ HF} \times \frac{1\ cm^3}{1.17\ g} \times \frac{\$0.75}{1000\ cm^3} = \$1.71$; no

81. $ClO^-(aq) + 2I^-(aq) + 2H^+(aq) \longrightarrow Cl^-(aq) + H_2O + I_2(s)$

$I_2(s) + 2S_2O_3^{2-}(aq) \longrightarrow 2I^-(aq) + S_4O_6^{2-}(aq)$

$n\ I_2 = (0.02500\ L)(0.0700\ mol\ S_2O_3^{2-}/L) \times \frac{1\ mol\ I_2}{2\ mol\ S_2O_3^{2-}}$

$= 8.75 \times 10^{-4}\ mol\ I_2$

mass NaClO = $8.75 \times 10^{-4}\ mol\ I_2 \times \frac{1\ mol\ ClO^-}{1\ mol\ I_2} \times \frac{74.44\ g\ NaClO}{1\ mol\ ClO^-}$

$= 0.0651\ g$

mass % = $\frac{0.0651\ g}{5.00\ g} \times 100\% = 1.30\%$

82. $NaN_3(s) \longrightarrow Na(s) + 3/2\ N_2(g)$

Assume T = 25°C; minimum P = 1.00 atm

$n\ N_2 = \frac{(20.0\ L)(1.00\ atm)}{(0.0821\ L\cdot atm/mol\cdot K)(298\ K)} = 0.818\ mol$

mass $NaN_3 = 0.818\ mol\ N_2 \times \frac{1\ mol\ NaN_3}{1.5\ mol\ N_2} \times \frac{65.02\ g\ NaN_3}{1\ mol\ NaN_3}$

$= 35\ g\ NaN_3$

CHAPTER 21
Organic Chemistry

LECTURE NOTES

This chapter is intended to give the student some idea of what organic chemistry is all about. We have tried to relate the material to topics with which students are familiar; the uses of acetylene, the properties of alcohol, the making of soap. We have deliberately avoided the encyclopedic approach, mentioning only a few functional groups with selected examples of compounds within each group. The nomenclature of alkanes is covered in some detail; alkenes, alkynes and aromatic hydrocarbons are treated more briefly.

The material in this chapter will require between two and three lectures depending upon how much time you choose to spend on nomenclature and polymers (Section 21.5). If you delete the latter and don't dwell on the formed, 2 lectures will suffice.

LECTURE 1

I <u>Hydrocarbons</u>

A. Alkanes: all single bonds; CH_4, C_2H_6, C_3H_8, - -

1. Structural isomerism - same molecular formula but differ in arrangement of atoms. Two isomers of C_4H_{10}, three isomers of C_5H_{12}. Isomers of $C_3H_6Cl_2$:

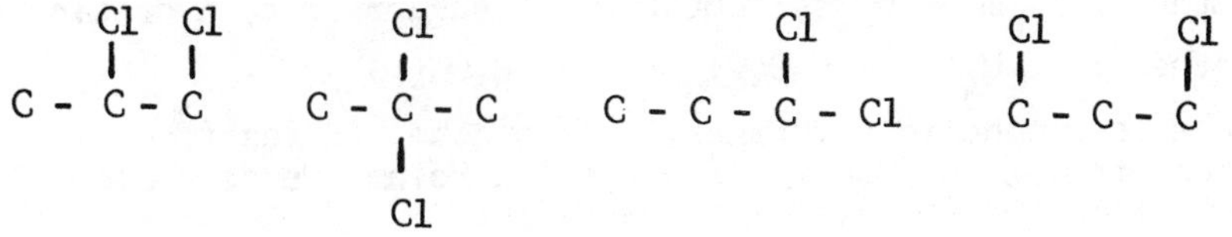

```
    Cl  Cl        Cl               Cl        Cl     Cl
    |   |         |                |         |      |
C - C - C     C - C - C    C - C - C - Cl    C - C - C
                  |
                  Cl
```

2. Nomenclature

a. Straight-chain: methane, ethane, propane, - - -

b. Branched-chain
suffix: find longest continuous carbon chain
prefix: methyl, ethyl, etc. Use number to indicate where branching occurs

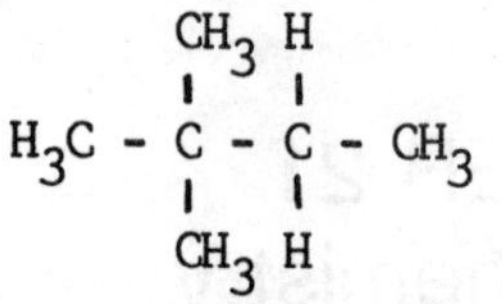

```
       CH3 H
        |  |
H3C - C - C - CH3
        |  |
       CH3 H
```

2,2-dimethylbutane

B. Alkenes: one double bond; C_2H_4, C_3H_6

Geometric isomerism; cis and trans

```
           Cl        H   Cl        H   Cl          Cl
             \      /      \      /      \        /
C2H2Cl2:      C = C         C = C         C = C
             /      \      /      \      /        \
           Cl        H   H         Cl  H           H
              (1)             (2)            (3)
```

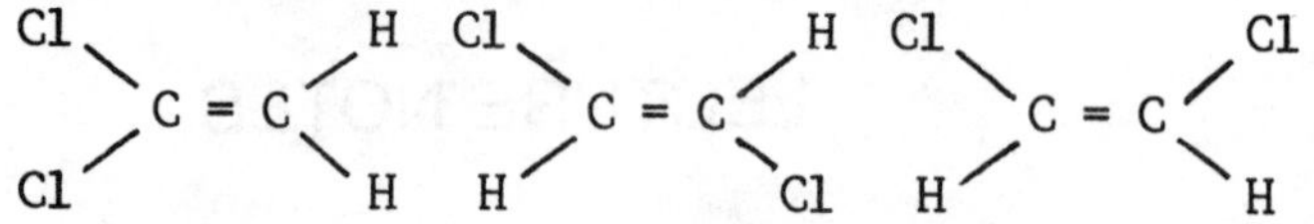

1 and 2 are structural isomers; 2 and 3 are geometric

LECTURE 2

I Hydrocarbons

A. Alkynes derivatives of acetylene, C_2H_2

B. Aromatic derivatives of benzene

benzene:  naphthalene:

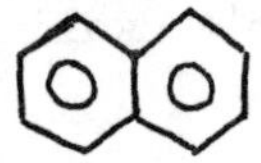

interpretation of "circle" structures

Nomenclature of benzene derivatives (ortho, meta, para)

II Functional Groups

A. Alcohols: -OH group (CH_3OH, C_2H_5OH, two isomers of C_3H_7OH)
Ethanol is made by fermentation of sugars or by hydration of ethene: $C_2H_4(g) + H_2O(g) \longrightarrow C_2H_5OH(l)$
Use in alcoholic beverages: % of alcohol varies from 4% in beer to 40% in brandy. Proof = 2 x Volume Percent alcohol

B. Acids:

```
- C - OH
  ||
  O
```

Formic acid: $HCOOH$ Acetic acid CH_3COOH

Treatment with base gives salt:

$CH_3COOH(aq) + OH^-(aq) \longrightarrow CH_3COO^-(aq) + H_2O$

Soaps are sodium salts of long-chain fatty acids:

$CH_3(CH_2)_{16}COO^-, Na^+$ sodium stearate, found in soap

C. Esters; formed by reaction of alcohol with carboxylic acid

$$CH_3-\underset{\underset{O}{\parallel}}{C}-OH + HO-CH_3 \longrightarrow CH_3-\underset{\underset{O}{\parallel}}{C}-O-CH_3$$

acetic acid methanol methyl acetate

Fats are esters of long-chain carboxylic acids with glycerol

$$\begin{array}{l} R_1-\underset{\underset{O}{\parallel}}{C}-O-CH_2 \\ \qquad\qquad\quad | \\ R_2-\underset{\underset{O}{\parallel}}{C}-O-CH \\ \qquad\qquad\quad | \\ R_3-C-O-CH_2 \end{array}$$

If R group contains a double bond, fat is unsaturated

LECTURE 2½

I Polymers

A. Addition polymers Alkene or derivative of alkene adds to itself to form a long-chain polymer.

Polyethylene (formed from C_2H_4)

$$\begin{array}{ccccccccccccccc} H & & H & & H & & H & & H & & H & & H & & H \\ | & & | & & | & & | & & | & & | & & | & & | \\ C & - & C & - & C & - & C & - & C & - & C & - & C & - & C \\ | & & | & & | & & | & & | & & | & & | & & | \\ H & & H & & H & & H & & H & & H & & H & & H \end{array}$$

may contain 2000 or more C_2H_4 units.

Polyvinyl chloride: made from C_2H_3Cl:

$$\begin{array}{ccccccccccccccc} H & & H & & H & & H & & H & & H & & H & & H \\ | & & | & & | & & | & & | & & | & & | & & | \\ C & - & C & - & C & - & C & - & C & - & C & - & C & - & C \\ | & & | & & | & & | & & | & & | & & | & & | \\ H & & Cl & & H & & Cl & & H & & Cl & & H & & Cl \end{array}$$

"head-to-tail" polymer

B. Condensation polymers

1. Polyesters: made from dicarboxylic acid + dialcohol

$$HO-CH_2-CH_2-OH + HOOC-C_6H_4-COOH$$

$$-O-CH_2-CH_2-O-\underset{\underset{O}{\parallel}}{C}-C_6H_4-\underset{\underset{O}{\parallel}}{C}-O- + H_2O$$

dacron

2. Polyamides: made from dicarboxylic acid and diamine

$$H_2N - (CH_2)_6 - NH_2 \quad + \quad HOOC - (CH_2)_4 - COOH$$

$$- \underset{\underset{H}{|}}{N} - (CH_2)_6 - \underset{\underset{H}{|}}{N} - \underset{\underset{O}{\|}}{C} - (CH_2)_4 - \underset{\underset{O}{\|}}{C} - \quad +$$

nylon

DEMONSTRATIONS

1. Acetylene from calcium carbide: Test. Dem. 47; J. Chem. Educ. 64 444 (1987)
2. Fermentation to ethanol: Test. Dem. 47
3. Formation of esters: Test. Dem. 48
4. Urea-formaldehyde polymer: Shak. 1 227

*5. Preparation of nylon: Test. Dem. 136, 164; J. Chem. Educ. 56 409 (1979); Shak. 1 213

*6. Optical activity: Test. Dem. 224; J. Chem. Educ. 53 506 (1976), 55 319 (1978); Shak. 3 386

7. Chemiluminescence: Shak. 1 133-200

*See also Shakhashiri Videotapes, Demonstrations 15, 43

QUIZZES

Quiz 1

1. Write the structural formula of
   a. two different alcohols containing three carbon atoms.
   b. the ester formed when acetic acid reacts with methanol.
   c. an alkyne with three carbon atoms.
   d. the addition polymer formed by propylene, C_3H_6.
   e. the condensation polymer formed by $HOOC - (CH_2)_5 - COOH$ and ethylene glycol, $HO - CH_2 - CH_2 - OH$.

Quiz 2

1. Write the structural formula of
   a. the acid and alcohol from which the following ester is made:

```
           H          H
           |          |
CH3 - C - C - O - C - CH3
           |   ||     |
          OH   O     CH3
```

b. all the isomers of C_4H_9Cl

2. Write the structural formulas of the head-to-tail, head-to head, and random addition polymers formed from $CH_2 = CHCN$.

Quiz 3

1. Draw all the isomers of the alkene C_4H_8.
2. Name the following

```
a. CH3 - CH - CH3      b. CH3 - COOH      c. CH3 - C = CH2
         |                                         |
        CH2                                        H
         |
        CH3
```

Quiz 4

1. Classify each of the following as an alcohol, carboxylic acid, and/or ester.

```
        H                        H
        |                        |
a. H - C - C - OH        b. H - C - C - O - CH3
        |   ||                   |   ||
       OH   O                   Cl   O

                 O
                 ||
          ╱ O - C - O - CH3
c.  (benzene)
          ╲ C - OH
            ||
            O
```

2. Draw the structures of all the isomers of $C_2H_2Cl_2$, a derivative of ethene.

Quiz 5

1. Draw the structural formulas of all the isomers of hexane, C_6H_{14} Name each of these isomers.

2. Draw the structure of a condensation polymer made from the two monomers:

```
        H                          CH3
        |                           |
CH3 - C - COOH      and     CH2Cl - C - COOH
        |                           |
       NH2                         NH2
```

Answers

Quiz 1 1. a. CH_3-CH_2-OH, CH_3-CHOH-CH_3 b. CH_3-C(=O)-OCH_3

```
                   O
                   ‖
b. CH3-C-OCH3
```

c. CH_3-C≡C-H

```
   H   H   H   H
   |   |   |   |
d. C - C - C - C
   |   |   |   |
   CH3 H   CH3 H
```

```
e. - C - (CH2)5 - C - O - CH2 - CH2 - O -
     ‖            ‖
     O            O
```

```
                 H                 H
                 |                 |
Quiz 2 1. a. CH3 - C - C - OH,  CH3 - C - CH3
                 |   ‖                 |
                 OH  O                 OH
```

b. four structural isomers in which Cl replaces H of butane.

```
     H   H   H   H          H   H   H   H
     |   |   |   |          |   |   |   |
2. - C - C - C - C -      - C - C - C - C -
     |   |   |   |          |   |   |   |
     H   CN  H   CN         H   CN  CN  H
```

```
  H   H   H   H   H   H
  |   |   |   |   |   |
- C - C - C - C - C - C
  |   |   |   |   |   |
  H   CN  H   CN  CN  H
```

```
Quiz 3 1. C - C - C = C,   C - C = C - C,      C - C = C  (4 in all)
                                                   |
                           (cis and trans)         C
```

2. a. 2-methylbutane b. acetic acid (ethanoic acid)
c.propene

Quiz 4 1. a. acid, alcohol b. ester c. acid, ester

```
   Cl       H   Cl       H   Cl       Cl
     \     /      \     /      \     /
2.    C = C        C = C        C = C
     /     \      /     \      /     \
   Cl       H   H        Cl  H        H
```

Quiz 5 1. five structural isomers; hexane, 2-methylpentane, 3-methylpentane, 2,3-dimethylbutane, 2,2-dimethylbutane

```
          H            CH2Cl
          |            |
2.  - N - C - C - N - C - C -
      |   |   ||  |   |   ||
      H  CH3  O   H  CH3  O
```

PROBLEMS

1. a. 2-metnyloutane b. 3-methylpentane
   c. 2,3-dimethylpentane d. 2,3,5-trimethylhexane

```
                                   C
                                   |
3. a. C - C - C - C - C    b. C - C - C - C
              |                    |
              C                    C
              |
              C

   c.  C - C - C - C - C - C - C     d. C - C - C - C - C
           |   |                            |   |
           C   C                            C   C
               |
               C

5. a.  C - C - C - C - C - C - C - C    4-isopropyloctane
                       |
                   C - C - C

   b.  C - C - C          2-methylbutane
           |
           C
           |
           C

   c.  C - C - C          2-methylbutane
       |   |
       C   C
```

7. a. 2-methyl-1-propene b. 2,3-dimethyl-2-butene
   c. 2-pentene d. 2-methyl-1-butene

9. a. m-dibromobenzene b. o-dinitrobenzene c. p-diiodobenzene

11. a. methanol, methyl alcohol b. ethanoic acid, acetic acid
    c. 1-propanol, propyl alcohol

```
13. a. H - C - O - CH3     b. CH3 - CH2 - C - O - CH2 - CH3
           ||                             ||
           O                              O
```

```
                                                                   CH3
                                                                   |
c. CH3 - C - O - CH2 - CH2 - CH3          d. CH3 - C - O - C - CH3
         ||                                        ||      |
         O                                         O       H
```

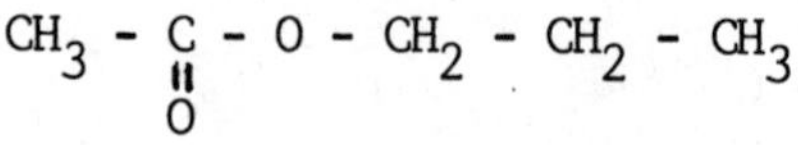

```
                                            C                C
                                            |                |
15.  C - C - C - C - C - C;   C - C - C - C - C;   C - C - C - C - C;

          C                  C   C
          |                  |   |
     C - C - C - C;     C - C - C - C
          |
          C
```

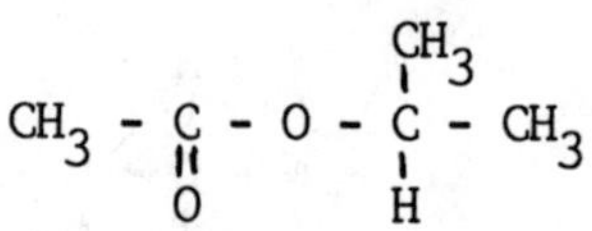

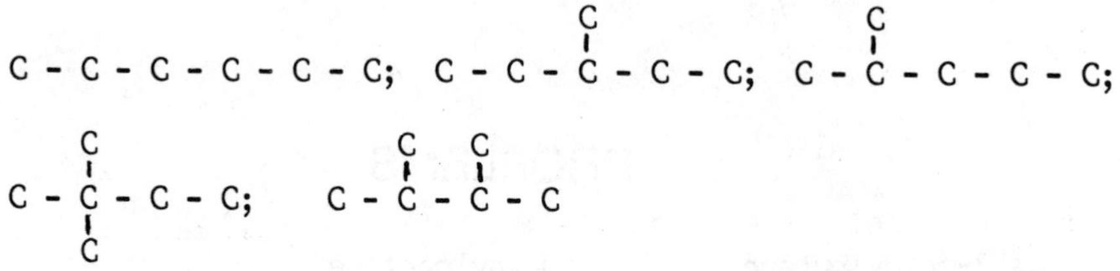

```
                                                       Cl
                                                       |
17.  C - C - C - C - Cl;   C - C - C - C;      C - C - C ;
                                   |                   |
                                   Cl                  C
          C
          |
     C - C - C - Cl
```

```
                                                      Cl
19.           -Cl                 -Cl
              -Br
                                   |
                                   Br                 Br
```

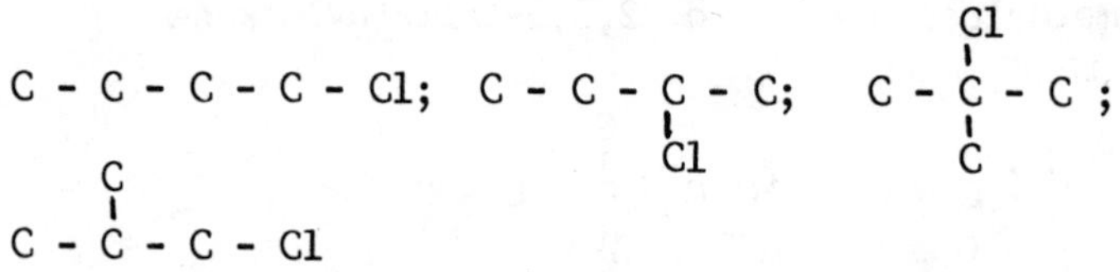

```
                                                     C
                                                     |
21.  C = C - C - C - C;   C - C = C - C - C;   C = C - C - C;

          C               C
          |               |
     C - C = C - C;   C - C - C = C
```

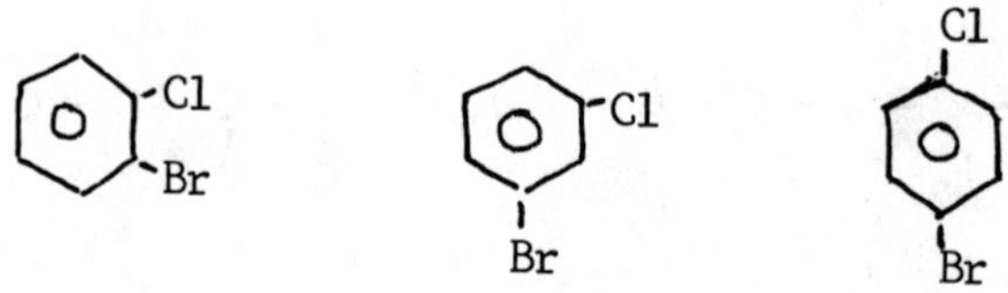

```
23.  C - C - C - C - C - OH;   C - C - C - C - C;   C - C - C - C -C
                                               |              |
                                               OH             OH

              C                   C                  C
              |                   |                  |
     HO - C - C - C - C;     C - C - C - C;     C - C - C - C;
                                  |                      |
                                  OH                     OH

          C                   C
          |                   |
     C - C - C - C ;     C - C - C - OH
                 |            |
                 OH           C
```

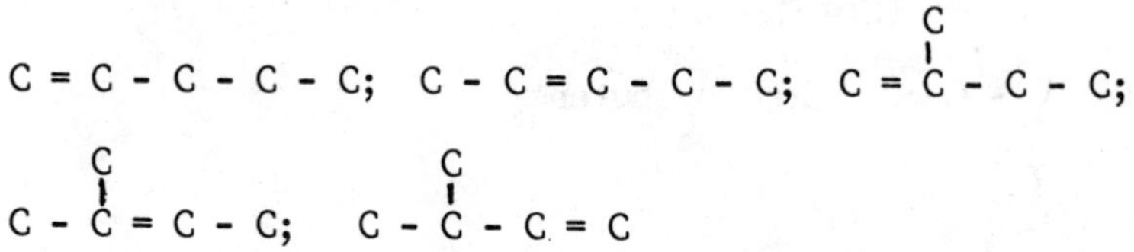

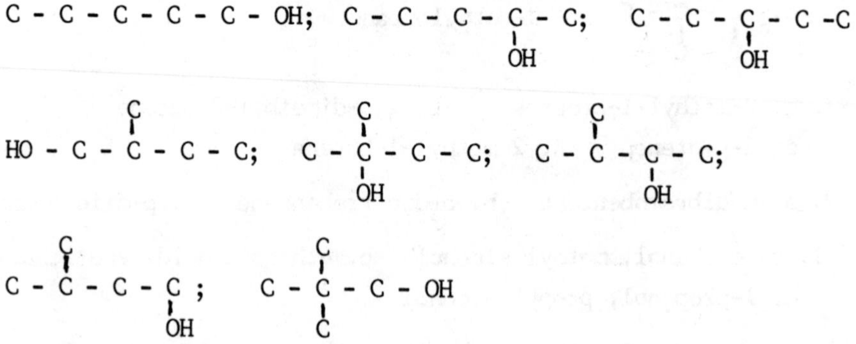

25. $CH_3(H)C=C(H)C_2H_5$ and $CH_3(H)C=C(C_2H_5)H$

$$\begin{matrix} CH_3 & & & & H \\ & \diagdown & & \diagup & \\ & & C = C & & \\ & \diagup & & \diagdown & \\ H & & & & C_2H_5 \end{matrix} \quad \text{and} \quad \begin{matrix} CH_3 & & & & C_2H_5 \\ & \diagdown & & \diagup & \\ & & C = C & & \\ & \diagup & & \diagdown & \\ H & & & & H \end{matrix}$$

27.

$$\begin{matrix} HOOC & & & & COOH \\ & \diagdown & & \diagup & \\ & & C = C & & \\ & \diagup & & \diagdown & \\ H & & & & H \end{matrix} \qquad \begin{matrix} HOOC & & & & H \\ & \diagdown & & \diagup & \\ & & C = C & & \\ & \diagup & & \diagdown & \\ H & & & & COOH \end{matrix}$$

malic fumaric

29. a. alcohol b. ester, acid c. alcohol, acid

31. a. $C - \underset{\displaystyle OH}{\underset{|}{C}} - C - C$ b. $C - C - C - C - COOH$

c. $CH_3 - \overset{\displaystyle H}{\overset{|}{\underset{\displaystyle C_2H_5}{\underset{|}{C}}}} - O - \underset{\displaystyle O}{\underset{\|}{C}} - C_4H_9$

33. a. $- \overset{\displaystyle Cl}{\overset{|}{\underset{\displaystyle Cl}{\underset{|}{C}}}} - \overset{\displaystyle Cl}{\overset{|}{\underset{\displaystyle Cl}{\underset{|}{C}}}} - \overset{\displaystyle Cl}{\overset{|}{\underset{\displaystyle Cl}{\underset{|}{C}}}} - \overset{\displaystyle Cl}{\overset{|}{\underset{\displaystyle Cl}{\underset{|}{C}}}} -$

b. $3.2 \times 10^3 \times 165.82$ g/mol $= 5.3 \times 10^5$ g/mol

c. CCl_2 C: $\frac{12.01}{82.91} \times 100\% = 14.49$; Cl = 85.51%

35. a. $- \overset{\displaystyle H}{\overset{|}{\underset{\displaystyle H}{\underset{|}{C}}}} - \overset{\displaystyle H}{\overset{|}{\underset{\displaystyle CN}{\underset{|}{C}}}} - \overset{\displaystyle H}{\overset{|}{\underset{\displaystyle H}{\underset{|}{C}}}} - \overset{\displaystyle H}{\overset{|}{\underset{\displaystyle CN}{\underset{|}{C}}}} -$ b. $- \overset{\displaystyle H}{\overset{|}{\underset{\displaystyle H}{\underset{|}{C}}}} - \overset{\displaystyle H}{\overset{|}{\underset{\displaystyle CN}{\underset{|}{C}}}} - \overset{\displaystyle H}{\overset{|}{\underset{\displaystyle CN}{\underset{|}{C}}}} - \overset{\displaystyle H}{\overset{|}{\underset{\displaystyle H}{\underset{|}{C}}}} -$

37. $H_2C = CHCl$ and $H_2C = C(H)(C_6H_5)$

39. $O - \underset{\displaystyle O}{\underset{\|}{C}} - \underset{\displaystyle O}{\underset{\|}{C}} - O - CH_2 - CH_2 - O -$

41. $H_2N - CH_2 - \underset{\displaystyle O}{\underset{\|}{C}} - OH$

```
           Br                          C
           |                           |
43. a. C - C - C - C    yes     b. C - C - C     no
           |
           Cl

                C                           C       C
                |                           |       |
    c.  C - C - C - C - OH    no   d.  C - C - C - C - C    no
                |                           |
                C                           C
```

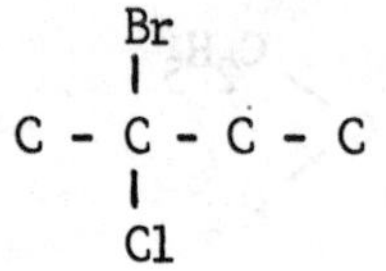

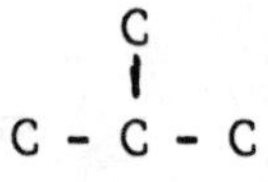

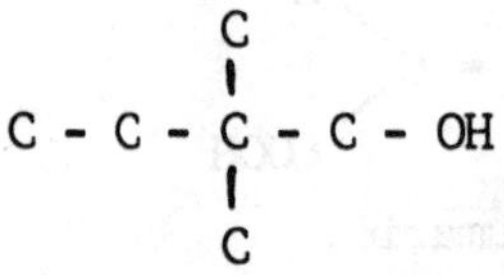

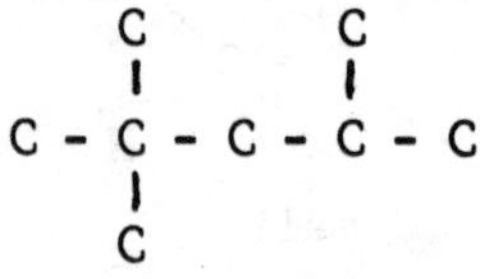

45. a. 2nd and 3rd from left b. none c. 3rd from left

47. a. 109.5° b. 109.5°, 120° c. 109.5°, 180°

49. three pairs of electrons spread around ring

51. a. reaction of CO with H_2 b. hydration of ethene
    c. oxidation of ethanol
    d. reaction of ethanol with acetic acid

53. a. no multiple bonds
    b. sodium salt of long-chain carboxylic acid
    c. twice the volume percent of alcohol
    d. ethanol with additives that make it unpalatable

```
           Br  H             H
           |   |             |
55. a. H - C = C - H    b. H - N - CH2 - COOH
```

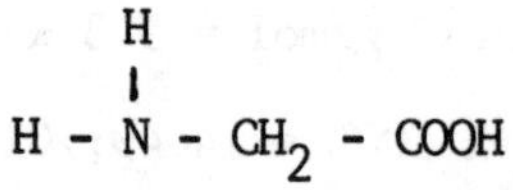

    c. HOOC - $(CH_2)_4$ - COOH and H_2N - $(CH_2)_6$ - NH_2

56. $C_{27}H_{46}O$

```
                                                          OH
                                                          |
57. C - C - C - C - C - C - OH;  C - C - C - C - C - C ;

                 OH                   C
                 |                    |
    C - C - C - C - C - C ;  HO - C - C - C - C - C

        OH                  OH                    OH
        |                   |                     |
    C - C - C - C - C;  C - C - C - C - C ;  C - C - C - C - C
        |                   |                     |
        C                   C                     C

    C - C - C - C - C - OH ;  HO - C - C - C - C - C
        |                                  |
        C                                  C
```

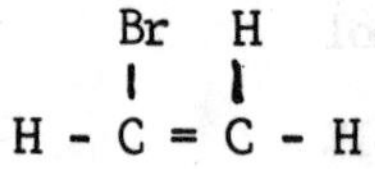

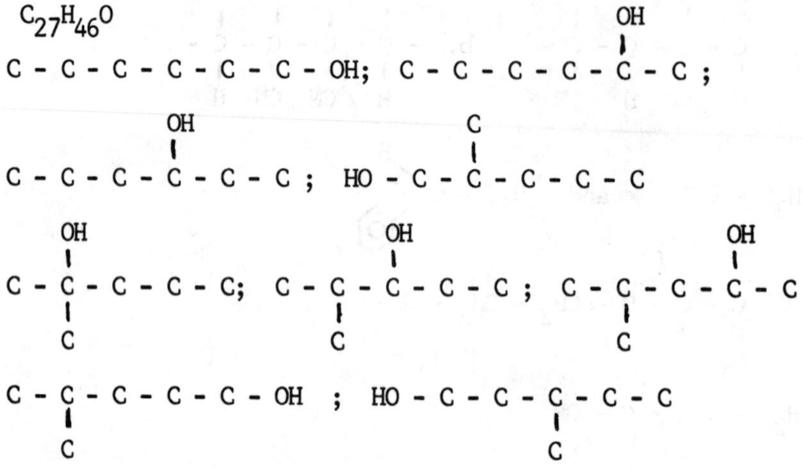

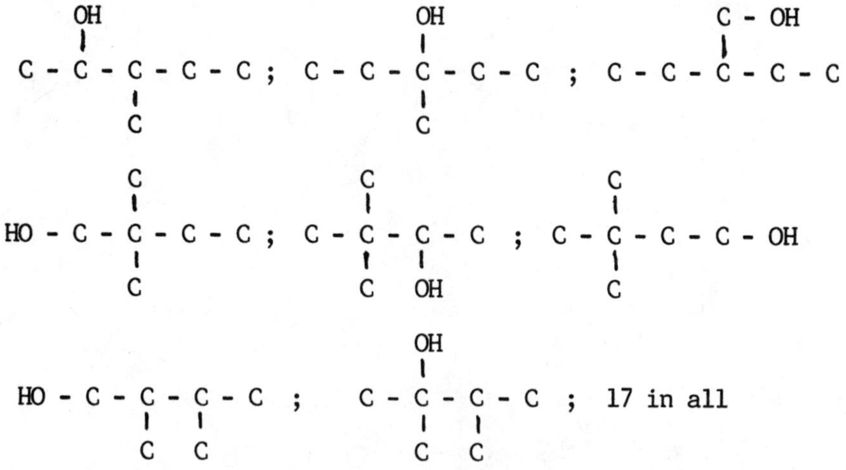

```
  OH                      OH                      C - OH
  |                       |                       |
C - C - C - C - C ;  C - C - C - C - C  ;  C - C - C - C - C
        |                 |
        C                 C

        C                 C                  C
        |                 |                  |
HO - C - C - C - C ;  C - C - C - C  ;  C - C - C - C - OH
        |                 |   |              |
        C                 C   OH             C

                          OH
                          |
HO - C - C - C - C ;  C - C - C - C  ;  17 in all
         |   |            |   |
         C   C            C   C
```

58. $C_3H_8(g) + 5\ O_2(g) \longrightarrow 3CO_2(g) + 4H_2O(l)$

$\Delta H° = 3(-393.5\text{ kJ}) + 4(-285.8\text{ kJ}) + 103.8\text{ kJ} = -2219.9\text{ kJ}$

$$q_{water} = 1.00\text{ qt} \times \frac{1\text{ L}}{1.057\text{ qt}} \times \frac{1000\text{ g}}{1\text{ L}} \times 4.18\ \frac{\text{J}}{\text{g}\cdot°\text{C}} \times 75.0°\text{C}$$

$$= 2.97 \times 10^5\text{ J} = 297\text{ kJ}$$

$$\text{mass propane} = 297\text{ kJ} \times \frac{44.09\text{ g}}{2219.9\text{ kJ}} = 6\text{ g}$$

(neglecting the heat capacity of the saucepan and assuming liquid is produced)

```
59. a.          H                (benzene ring)              H
                |                                            |
     O - CH2 - C - CH2 - O - C               C - O - CH2 - C - CH2 - O -
                |              ||             ||              |
                OH             O              O               OH

    b.            H               (benzene ring)              H
                  |                                           |
       - O - CH2 - C - CH2 - O - C             C - O - CH2 - C - CH2 - O -
                  |              ||            ||             |
                  O              O             O              O
                  |                                           |
                  C = O                                       C = O
      (benzene ring)                             (benzene ring)
                  C = O                                       C = O
                  |                                           |
                  O              O             O              O
                  |              ||            ||             |
       - O - CH2 - C - CH2 - O - C             C - CH2 - O - C - CH2 - O -
                  |              (benzene ring)               |
                  H                                           H
```

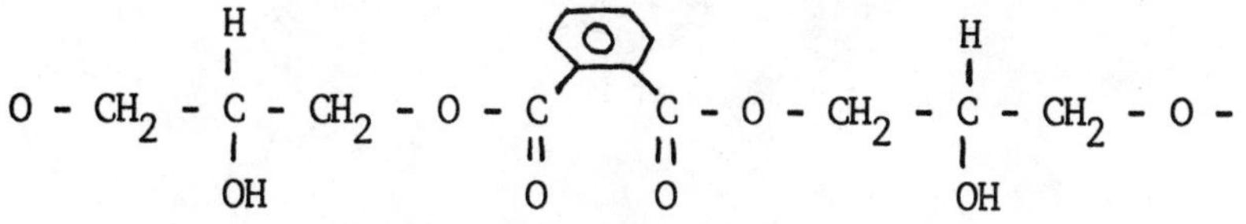

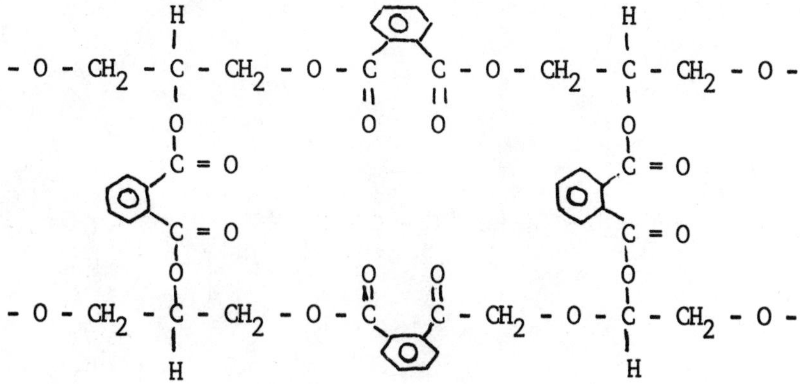